NOW YOU'RE TALKING

ALSO BY TREVOR COX

The Sound Book

NOW YOU'RE TALKING

Human Conversation from the Neanderthals to Artificial Intelligence

Trevor Cox

COUNTERPOINT
Berkeley, California

To Deborah, Jenny, Michael, Nathan and Peter

Now You're Talking

First published in 2018 by The Bodley Head, an imprint of Vintage.
Vintage is part of the Penguin Random House group of companies.

First Counterpoint hardcover edition: 2018

Library of Congress Cataloging-in-Publication Data
Names: Cox, Trevor J., author
Title: Now you're talking : human conversation from the Neanderthals to artificial intelligence / Trevor Cox.
Description: First hardcover edition. | Berkeley, California : Counterpoint, 2018 | Includes bibliographical references and index.
Identifiers: LCCN 2018015090 | ISBN 9781640090798
Subjects: LCSH: Oral communication—History.
Classification: LCC P95 .C69 2018 | DDC 302.2/242—dc23
LC record available at https://lccn.loc.gov/2018015090

Jacket designed by Jaya Miceli

COUNTERPOINT
2560 Ninth Street, Suite 318
Berkeley, CA 94710
www.counterpointpress.com

Printed in the United States of America
Distributed by Publishers Group West

1 3 5 7 9 10 8 6 4 2

Contents

Introduction

> 'I regard this invention as an epoch in the history of science … a vote of thanks … for the great pleasure we have had … in hearing Mr Edison's instrument.'[1]
>
> 'Your marvelous invention has so occupied my brain that I can hardly collect my thoughts to carry on my work. The results are far reaching (in science), its capabilities are immense.'[2]

In December 1877, Thomas Edison made history by recording 'Mary Had a Little Lamb' on his phonograph and playing it back. This was not just 'an epoch in the history of science', it was a revolution for the human voice. Before then, hearing someone talk was exclusively a live experience: you had to be listening as the sounds emerged from the speaker's mouth. We can read the text of great speeches that predate the phonograph, like Abraham Lincoln's Gettysburg Address, but how exactly the president delivered the lines is lost forever. The phonograph captured the way things are said, and this can be just as important as the words themselves.

When someone says 'I'm all right', the tone of their voice might in fact tell you they are *not* all right.

Voice lies at the heart of our identity. It takes only a few words to recognise a friend or loved one when they call you on the phone – this is an astonishing ability. If a stranger calls, we immediately start picking up on accent and intonation and make assumptions about the talker's class, background and education. We also make inferences about their age, stature and personality, although these are often wrong and compromised by preconceptions and prejudice. We adapt our speech to alter how we are perceived. We are oral chameleons, subconsciously turning our accent up in our home town or down elsewhere to fit in. Our vocal character is less fixed than you might imagine.

How the voice shapes our sense of self is highlighted by the disconcerting experience of hearing a recording of our own speech. We spend a lifetime listening to a voice that appears more boomy than the one that others hear, because bone vibrations carry the sound internally from the larynx to the ear and boost the bass. A recording quickly reveals that the vocal identity we present to others does not match our own inner voice. Before Edison's invention we lived in blissful ignorance of this.

Humanity's 'oral history' can be split into three eras, with the phonograph marking an epoch. In the beginning we were like other animals, making simple vocalisations to influence others: calling to ward off rivals, warn of danger or attract a mate. The second era began with the emergence of human language, which facilitated huge feats of collective endeavour and allowed us to dominate the world. A good proportion of human speech was still aimed at influencing the thoughts and behaviour of others – whether that is a parent telling a toddler not to run into the road, or Henry V rallying the troops with a cry of 'Once more unto the breach' – but we also talk for pleasure and to amuse ourselves, engage with the world, and to declare our love. The arrival of technology like the phonograph, which marked the beginning of the third era, enabled individuals to reach out to large groups of people, sometimes with devastating effect. One German minister stated at the Nuremberg Tribunal that the Nazi dictatorship was the first 'which made complete use of all technical means for the domination of its own country. Through technological devices like the radio and the loudspeaker, eighty million people were deprived of independent thought.'[3] We are now at the dawn of an exciting new era where artificial intelligence (AI) means that we are starting to converse with computers. For better or worse, our exceptional abilities in verbal communication are becoming less unique, as we start sharing them with machines.

Now You're Talking is the story of how speaking and listening evolved, how we each develop these remarkable talents during our childhood, and how human communication is being changed by technology. Engaging in conversation seems a simple task because we are so adept at doing it. But in reality, speaking and listening are some of the most complex tasks our body and mind have to perform. Talking requires precise anatomical gymnastics, exquisitely masterminded by many different regions of the brain. Working out what someone is saying, and decoding cues about meaning and mood from the tone of the voice, is immensely complicated. These processes are normally hidden from view, but psychologists, neuroscientists and biologists are revealing more and more of what is going on. In today's world, many unmediated face-to-face conversations are being supplanted by talk that is transmitted and transformed by technology. The influence of technology is only going to increase as conversing with computers becomes commonplace. What secrets might we unwittingly reveal to our devices? How does AI listen and speak, and how will it change human speech in the future?

The phonograph is an exemplar of how technology has imposed itself on speaking and listening. It was first introduced to a British audience in February 1878. The demonstration took place in the Royal Institution, the venue where in the Victorian era the great and the good gathered to be entertained by the latest in science and engineering. The lecture theatre was packed as William H. Preece, chief engineer of the British Postal Telegraph Department, demonstrated a replica of Edison's invention. It had been hurriedly assembled over the previous week because of a delay in getting a phonograph shipped from America. Like Edison, Preece turned to a nursery rhyme to test the device, reciting 'Hey Diddle Diddle, the Cat and the Fiddle'. 'The words were distinctly heard, but the voice was a very faint and unearthly caricature,' reported the *London Weekly Graphic*. Using a children's nursery rhyme to showcase

Thomas Edison and his phonograph.[5]

a groundbreaking piece of engineering was a clever choice: the listener knew the lyrics so could subconsciously fill in words lost in the scratching noise as the needle danced across the tinfoil. The new invention was a sensational hit. 'A crowd collected round the table to see, speak to, and hear the phonograph,' wrote the *Graphic*, 'and the theatre was not cleared until eleven o'clock, when the gas was turned out: a very broad hint it was time to go.'[4]

The second tinfoil phonograph that Edison built arrived in England within a fortnight. It is not normally on public display, so I was privileged to see it up close when taking part in a BBC radio broadcast. At the right side there is a crank handle to rotate the central cylinder that is covered in tinfoil. At the left side is a large flywheel to smooth the motion. You speak into a simple funnel that focuses the sound onto a small membrane which then vibrates. Attached to the back of the membrane is a needle that carves out a spiral groove as the tinfoil spins around. It is ingeniously simple: air vibrations that make up the sound of the voice are transformed into oscillations of the needle, with a record of the needle's motion engraved into the foil as a corrugated groove. To reproduce the sound you follow the same path in reverse, with a pick-up needle riding along the groove, following the dips and bumps, creating vibrations of the membrane and then air molecules, which travel to the listener's ear.

Edison's phonograph is a museum piece that is no longer played, but while I was in the Royal Institution I recorded Alfred, Lord Tennyson's poem 'Come into the garden, Maud' on another machine. This verse was an obvious choice because Tennyson himself witnessed it being recited to the phonograph when the invention premiered at the Royal Institution. You have to get very close to the speaking horn and shout to make sure the indents are large enough, otherwise on replay the words get swamped by surface noise. On playback my voice sounded strangulated but the speech was clearly intelligible above the inevitable scratching noise.

Early phonograph demonstrations included playful experiments. A popular party piece at the time was to change the speed of the crank while replaying. One witness described hearing 'an angry old woman' when the cylinder was rotated too fast and 'a decrepi[t] old man with his mouth full of water' when the replay was slowed down.[6] The Beatles were famous for pioneering experiments in sound – superimposing voices, playing recordings backwards and at different speeds – and in the 1970s, religious groups were irked by supposed satanic lyrics being transmitted if you played songs like 'Stairway to Heaven' by Led Zeppelin backwards. But Edison got there first by reversing his recording of 'Mad dog! Mad dog! Mad dog!'

The effect of technology on the voice has been much more profound than merely allowing us to play around with speech recordings: it has changed how we speak and sing. I recently compared a historic phonograph recording of the actor Sir Henry Irving reciting 'Now is the winter of our discontent', to a modern rendition by David Morrissey. The nineteenth-century recording has Irving projecting his plummy voice using a vocal technique honed for the stage of large theatres. In contrast, the microphone frees Morrissey from having to talk very loudly, and he speaks the lines as though performing to a small gathering, with the breathy subtleties of his voice clearly audible. Changes to how we sing have been even more dramatic. You might compare early phonograph recordings of the operatic superstar Adelina Patti to a great modern singer such as Amy Winehouse. Patti's operatic sound has much purity and sweetness, whereas Winehouse's delivery expresses far more personality and soul. Patti had to precisely shape her vocal anatomy to produce a loud sound. With electronics taking the load, Winehouse had much more freedom for expression. Technology has given us the huge variety of voices heard in modern music.

Piegan Indian and ethnologist Frances Densmore in 1916.

Sound is naturally ephemeral, but that changed with recording. We now have a rich aural history of speech for scientists to analyse. This has revealed cultural changes, such as the lowering of women's voice pitch in recent decades or the replacement of the cockney twang with blended multicultural accents in the East End of London. The human voice would have morphed over humanity's entire history, but it is only now that changes can be directly observed. Scientists can even compare old and new recordings to hear how a lifetime of speaking and listening changes an individual's voice. Fortunately, our vocal anatomy fends off old age well; wrinkles and grey hairs appear long before our voices deteriorate.

Edison envisaged many uses for his invention, but most poignantly he wanted to record people before they died. The voice expresses personality more vividly than a picture because it is *alive*. As Edison predicted, 'for the purpose of preserving the sayings, the voices, and the last words of the dying member of the family – and of great men – the phonograph will undoubtedly outrank the photograph'.[7] While this prophecy missed the mark, people are now increasingly documenting loved one's voices via videos. Mobile phones and other gadgets mean that moving images and soundtracks are vying for attention with still photographs.

Nowadays, AI allows vocal memorials that simulate conversations with the dead. There is much talk of AI putting workers out of a job, but who would have thought that clairvoyants presiding over séances would face redundancy? (But presumably they already know that and are retraining.)

In 2015, the digital-magazine editor Roman Mazurenko was tragically killed by a speeding car in Moscow. His close friend, the tech entrepreneur Eugenia Kuyda, built a chatbot so she could talk to Roman one last time. Eugenia had thousands of text messages from her friend, and these were fed into a computer program that employed AI to create a bot that used Roman's turn of phrase. While Tennyson's words are permanently etched into wax, the Roman bot can reply with new phrases that never existed in the text messages.[8] Here is a typical exchange:

> Eugenia: How are you?
> Roman bot: I'm OK. A little down. I hope you aren't doing anything interesting without me?
> Eugenia: A lot is happening. Life is going on, but we miss you.
> Roman bot: I miss you too. I guess this is what we call love.[9]

What should we make of all this? Talking to a departed loved one, both out loud and in your head, is not unusual, but the thought of a machine providing answers is creepy. Roman's friends and family were divided by the technology: some liking it while others found it distasteful. Now imagine going one step further and using recordings to reconstruct Roman's voice. This is perfectly feasible: personalised artificial voices are becoming more and more common for those who have lost their voice due to conditions like motor neurone disease. If you feel uncomfortable with the thought of a memorial chatbot texting, imagine a macabre machine bringing a loved one's voice back to life. This raises many ethical questions, such as whether it is right to raid someone's digital footprint to create a semblance of immortality.

AI is poised to change our conversations fundamentally. For humans, speaking and listening are not just ways of passing on factual information. The phrase 'I love you' is loaded with connotations. It seems an unlikely phrase to say to a computer, but every day thousands of people profess their love to Alexa, the voice-activated home assistant from Amazon.[10] As we develop machines that comprehend and portray emotions, or even just mimic them convincingly, our relationship to these devices changes forever. We are not so far from the scenario depicted in the 2013 movie *Her*, where a lonely man falls in love with an artificially intelligent operating system called Samantha.

As technologies get better at conversing naturally, whose jobs will be on the line? In the early nineteenth century, the Luddites smashed the new machines of the Industrial Revolution that threatened their livelihood. As music recordings became more common in the early twentieth century, the composer John Philip Sousa feared that soon 'no one will be ready to submit himself to the ennobling discipline of learning music'.[11] In 2014, a production of Richard Wagner's Ring Cycle in Hartford, Connecticut was postponed following an outcry over the use of a computer instead of a pit orchestra.[12] If machines gain emotional agency, will we see Luddite luvvies storming the Globe Theatre to smash androids reciting Shakespeare? Could AI go even further, taking the place of the Bard and writing a play for the robots to perform?

In theatre there is a long tradition of using animals, ghosts or puppets to shed light on humanity. As computers start to converse with us, technology will also reveal much about ourselves. Contrast the struggle scientists are having with getting computers to listen and talk with the way that children naturally develop the ability. We think that doing complex sums is hard and having a conversation is easy. But when we try to get machines to do these tasks, it turns out that the sums are the easy bit. The human ability to have a conversation seems so simple, but it is truly remarkable.

While today speech and hearing are often entwined with technology, if we want to understand the human ability to converse we need to unpick what happened long before the invention of the phonograph. How did human speech evolve? Were Neanderthals able to chat with 'modern man', *Homo sapiens*? This hotly debated topic is the subject of the first chapter.

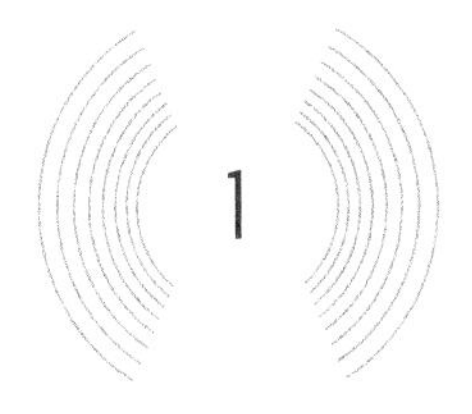

1

Evolution

'Language is the Rubicon which divides man from beast, and no animal will ever cross it', the Oxford professor Max Müller declared in 1861.[1] It is our ability to reason through language that separates humans from other animals. Or as Müller put it, 'No reason without speech, no speech without reason.'[2] The professor believed language was divine in origin, and was a passionate opponent of Darwin's theory of evolution by natural selection.[3] But he had a misplaced confidence in his ability to win the argument, claiming that 'the science of language will yet enable us to withstand the extreme theories of the Darwinians'. A decade later Darwin rose to the challenge and outlined how language might evolve by natural selection in his remarkable book *The Descent of Man*. But still the arguments rumbled on. Two years later, the Linguist Society of Paris banned discussion on the origins of language to reduce rampant speculation based on little hard evidence.

Language is what makes us human, so no wonder many great thinkers have come up with theories for how we came to speak. But peering back hundreds of thousands of years to work out whether one of our ancestors spoke a language is extremely challenging. Sound is ephemeral, disappearing as soon as it is made, so it's difficult to know what, if anything, our ancient ancestors might have said or heard. While fossil evidence plays a vital role in understanding many aspects of evolution, it is less useful for language: the brain does not fossilise, and neither does the vocal apparatus. Still, the paucity of evidence creates fertile ground for enticing theories to be developed and then hotly disputed. As the science writer Philip Ball wrote about similar arguments concerning the evolution of music, 'the stridency with which points of view are asserted seems to bear an inverse relation to the quantity and quality of supporting evidence'.[4] Even today, scientific papers and books are published that provoke the ire of other academics, who then write stinging rebukes. The field is fractious, an extreme example of how science develops through hypotheses being forensically analysed by peers, many of whom revel in finding flaws in a rival's pet idea. A recent critique in the scientific journal *Frontiers of Psychology* oozes contempt even in the title, 'Neanderthal language? Just-so stories take center stage'.

Take a step back from the controversy, and it is clear that modern science can do so much better than folk theories and speculation. As we shall see, scientists have come up with ingenious ways of probing evolutionary history. So while a definitive answer as to when verbal communication evolved may still be elusive, science can give a fascinating insight into how we developed this incredible ability.

While a verbal language needs both speech and hearing, it is talking that is usually coveted as something uniquely human. We don't seem to feel as threatened by the thought that animals might understand what we're saying. This might be one reason why the evolution of hearing is much less controversial than speech. Also, there is a more complete fossil record of the mammalian ear, which helps limit speculation.[5]

When our vertebrate ancestors (tetrapods) first left the sea about 350 million years ago, they were probably drawn out of the water to feast on land-dwelling invertebrates. Acanthostega is an example of one of these early tetrapods, looking like a squashed ugly eel with short stubby legs.[6] Tetrapods probably had both gills and lungs to allow them to breathe below and above water. They were only well adapted to hearing underwater, however. Their hearing anatomy evolved for aquatic life and would have been virtually useless when their head was in air. Sound waves are tiny fluctuations in pressure. Underwater these are carried by the water, whereas on land sound is vibrations of air molecules. Water and air are very different materials and consequently tetrapods would have struggled to pick up the tiny motions of air molecules. We have all experienced the converse case: human hearing is designed to work well in air, but stick your head underwater in a swimming pool and sounds become muffled.

Lungfish are the closest living relatives to the tetrapods and so studying them gives us some insight into the development of hearing. This is why Christian Christensen experimented on the species for his PhD at Aarhus University.[7] If lungfish were utterly deaf to aerial sound, he wondered, how did hearing evolve? For his experiments, he wrapped a lightly anaesthetised fish in damp paper towels and placed it on a hammock in the centre of a silent room, to ensure that the fish could only respond to sounds Christian played through his loudspeakers. The fish had electrodes placed on its head to allow him to monitor the fish's neurons.

Contrary to what Christian expected, it turns out that lungfish are not completely deaf. At low frequencies, below 200 Hz, the fish could pick up sounds above eighty-five decibels. Imagine an itinerant trombonist happening to wander by and play a loud low note in the room. Although the lungfish lacked a sensitive ear, it could still 'hear' the note: the sound would make the whole head of the fish vibrate and it was that motion that the brain can sense. 'It was a surprise that the lungfish, being completely unadapted to aerial hearing, were in fact able to hear airborne sound,' Christian told me. 'This tells us that even the early tetrapods and possibly also their aquatic ancestors may have been able to detect airborne sound.' For the tetrapods, this basic aerial hearing would have been too poor to be useful, however. Predators could easily creep up on them unheard – provided they were not playing the trombone. But while not of much immediate use, this rudimentary sensing provided a starting point for selective evolutionary pressures to improve upon.

Mammals have much more sensitive hearing than the early tetrapods because of many evolutionary adaptations. Sound is first amplified by the resonance of the ear canal and the concha, the small bowl-shaped depression of the outer ear. At most this gives about twenty decibels of gain, roughly equivalent to a quadrupling of perceived loudness. The second boost comes from the middle ear,

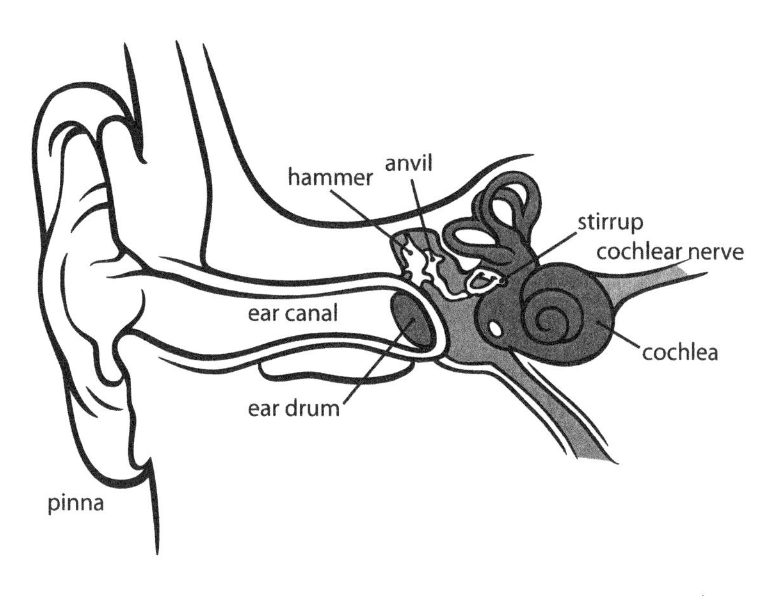

The human hearing system

which consists of the eardrum (tympanic membrane) and three tiny bones: malleus (hammer), incus (anvil) and stapes (stirrup), collectively known as the ossicles. Here, the tiny air movements of the sound waves are converted into the physical vibrations of body parts. Finally there is more amplification in the cochlea of the inner ear, as the vibrations get transformed into electrical impulses that then pass along the auditory nerve to the brain.

In studying the evolution of aerial hearing, the focus is usually on how the middle ear adapted to aid survival on land. The eardrum is a very lightweight membrane about nine millimetres wide. It collects sound across most of its surface, and at the threshold of hearing the eardrum moves by only a fraction of the diameter of a hydrogen atom. The eardrum is such a useful adaptation that it has evolved independently in mammals, reptiles and birds.[8] Next the sound moves the hammer and anvil, which act like a lever system to amplify the force that vibrates the stirrup bone. The boost from the middle ear mostly arises from the difference in size between the eardrum and the footplate of the stirrup, which then pushes on the entrance to the inner ear. A useful analogy is to imagine a centipede with a hundred legs, which for some reason is just balancing on six of them. Under each of the grounded legs there is a greater force than if all feet were on terra firma: the increase in force is by a factor of about seventeen (one hundred divided by six). This is the same amount of amplification that the incoming sound receives due to the force across the eardrum being concentrated over the smaller area of the stirrup's footplate. Overall, the middle ear amplifies sound by about thirty decibels – roughly the difference in loudness between talking normally and shouting.[9]

It would be tempting to simplify and tell the story of hearing evolution as a straightforward process where mammals' anatomy changed to amplify sound using the mechanisms above. But in reality, evolution pathways are usually more complicated. Structures develop usages that they were not originally designed for, something biologists call *exaptation*. Ironically, the scientist who first documented the development of the ear bones and gave a unique insight into hearing evolution, Karl Bogislaus Reichert, was not a fan of Charles Darwin's work.

Reichert was a nineteenth-century German anatomist. Photos show him with a fine mane of swept-back hair and oval, metal-rimmed glasses; on some he sports an impressive Van Dyck beard. Although Reichert made one of the most important discoveries in vertebrate biology he is little known nowadays and has even been cruelly described as 'solid but not intellectually outstanding'.[10] Early in his scientific career, Reichert dissected pig embryos and realised that two of the ossicles, the hammer and anvil, begin as cartilage attached to the back of the embryonic jaw. As the embryo develops further the cartilage ossifies, shrinks and moves away from the jaw to form two of the middle ear bones. As Reichert remarked in 1837: 'Rarely does one find a part of the animal organism in which changes from its original formation are so clearly evident as the middle ear bones of mammals.' Yet two decades on and Reichert was reluctant to extrapolate from his observations into accepting that Darwin's theory of evolution by natural selection could explain what he had seen with his own microscope.

So how does studying the development of a modern-day animal from embryo to adult enlighten our understanding of evolution that happened millions of years ago? I spoke to an expert in evolutionary developmental biology, Vera Weisbecker, from the University of Queensland. Ancestral characteristics can be preserved in an organism's development, she told me. 'Evolution adds new processes onto old processes. We look not unlike apes because of our ancestry, and ... a lot of our developmental trajectory is ape with minor alterations.' This means that when we look at our development we see some of our evolutionary past. And this is why Reichert's observation on the development of pig embryos was so important: it revealed that mammals have a template for a reptilian jaw, from which the middle ear bones develop and evolve.

But Reichert did not get it at the time. He firmly believed that 'The idea that the embryo of higher animals, in its individual development, passes through those of the lower animal world … is not supported by the current state of science.' This meant that though he had made extraordinary observations as a student, Reichert was left behind when the theory of evolution transformed biology. As one rival, Ernst Haeckel, cruelly commented, 'I have clearly demonstrated the utter worthlessness of Reichert's assertions, and the perversity of his misrepresentations … On reading [Reichert], we seem to have retrograded about half a century.'[11]

Reichert's observations on pigs are just one insight into how the large jaw joints of our ancient reptilian ancestors evolved to form delicate bones in our middle ear. The evolutionary story begins a little under 300 million years ago with the synapsids, a group of creatures that would later evolve into mammals. An early synapsid was the dimetrodon, which with a large sail-like structure sticking out of its back looks more like a dinosaur than a mammal. As the synapsids evolved into mammals over roughly 80 million years, the jaw joint changed several times.[12] This included two bones from the jaw joint shrinking and migrating towards the ear to form two of the ossicles.

Key evidence is provided by the fossilised remains of yanoconodon, a tiny mammal just over ten centimetres long, which were found in the Yan Mountains of China.[13] Dating from 125 million years ago, the Mesozoic era when dinosaurs roamed the earth, yanoconodon probably lived in the undergrowth eating insects and worms. The fossil appears to show a transitional structure before the detachment of the ossicles from the jaw. So the yanoconodon would have been able to hear high-frequency airborne sounds, while still having the reptilian ability to pick up ground vibrations through the jaw.

It would be a neat story if yanoconodon possessed the intermediate structure between reptilian and mammalian hearing. But with the fossil record being so sparse, this could be oversimplifying the evolutionary pathway. The bony connection might just be a specialisation in yanoconodon that was not passed on. Maybe the evolutionary pathway involves another mammal that is missing from the fossil record? Unfortunately the remains of middle ear bones are rarely found *in situ*. During decomposition and fossilisation, a skeleton often undergoes a lot of abuse – being moved by rivers, attacked by scavengers or simply trampled and crushed.[14] No wonder the tiny ossicles are often missing.

To supplement the sparse fossil record, scientists turn to evolutionary development biology, known as 'evo-devo' for short, where embryonic development is examined to better understand how an organism evolved. This is what had motivated my call to Vera Weisbecker, because she had just published a paper showing that one common evo-devo story was wrong. As she explained, over-interpretation of data from development can lead evolutionary science down the wrong track; done correctly, however, evo-devo is very powerful. Vera had been working on the development of marsupials. Here, a similar transition from jaw to ear happens in neonates, thereby mimicking what happened during mammalian evolution. In the first few weeks following birth, marsupials suckle with a jaw joint formed between the anvil and hammer bones. Over the following weeks, however, the jaw joint reconfigures and these bones migrate and become part of the middle ear.

Vera and her colleagues collected many samples of marsupial juveniles at different ages.* Using a CT scan that takes many

* Vera wanted me to make it clear that the marsupials were not killed especially for this research; she used samples that had been collected for other studies.

X-ray images, they examined when the middle ear bones became separated from the jaw, as well as measuring the size of the bones. Small ossicles are needed to create good hearing sensitivity because high-frequency sound waves would struggle to move larger bones. If hearing was the first driver for the creation of tiny ossicles, Vera expected the shrinking of the bones to be the trigger for the hammer and anvil detaching from the jaw. But in fact, the two ossicles detach from the jaw *first*, and *then* shrink. This implies that there is another evolutionary driver behind the detachment that has nothing to do with hearing. Judging by when this happens in a marsupial's development, it is probably connected with the emergence of the rear molars.

If mammal evolution follows the same path as marsupial development, this would mean that two of our middle ear bones first formed to help with eating and not hearing. One theory has it that the driver was changes in diet and the need to be able to crush seeds. Only later did the detached bones shrink, change function and get used for hearing. Here we have a perfect example of exaptation.

One strange feature of human hearing is that it extends over a much wider bandwidth than needed for communication. For a young adult, hearing extends to about 20,000 Hz, but we only need the bottom fifth of that range to understand speech. (This is something that phone companies exploit to reduce the bandwidth of calls.) What selective pressures might explain our high-frequency acuity? Millions of years ago mammals were small animals scurrying around in the undergrowth trying to avoid dinosaurs. They needed high-frequency hearing to pick up on each other's squeaks. But when mammals grew larger and humans evolved, why did the hearing range not reduce? According to Rickye and Henry Heffner from the Department of Psychology at the University of Toledo, this is because high frequencies are needed to work out where sound is coming from, and this provided a selective evolution pressure that shaped hearing.

Locating sounds is vital for animals, whether you are a predator looking for food or a vulnerable creature trying to avoid becoming someone else's lunch.[15] A couple of the techniques mammals can use to locate sound helps explain why we have two ears, because they involve comparing what we hear in each of them. When sound comes from the front, the path it travels to both ears is the same, as the head is symmetrical and so the signals going to the brain up the left and right auditory nerves will be identical. But for sound coming from the side, there are differences. The far ear receives it later because the sound takes extra time to get there; this location cue is most useful at lower frequencies. At high frequencies the sound is also quieter at the far ear as it has to bend around the head; this provides a second clue as to where the sound is coming from.[16]

The quality of these two cues depends on how far apart the ears are. For a big mammal, say an elephant, the sound has to bend around a large head resulting in a bigger time delay between the two ear signals and the furthest ear receiving an even quieter sound. This means elephants can do useful localisation even with low-frequency sound. In contrast, a small mammal such as a shrew has to use higher frequencies.*

So you'd think the ability to localise sound would be strongly correlated with head size, but it isn't. Place someone in front of an arc of loudspeakers, and a person can tell which of the speakers is making sound with startling accuracy. A human can pinpoint a source to within one or two degrees for sound coming from straight in front. Carry out a similar experiment with a

* There is a strong correlation between the highest frequency a mammal can hear and its head size.

horse and you would find much poorer localisation, something like twenty-five-degree accuracy. The width of a horse's head is similar to that of a human, so it is receiving similarly strong cues for locating sounds. But for some reason the evolutionary pressure for accurate aural localisation was stronger in humans than horses.

'Horses and cattle are horrible localisers, I mean just terrible,' was Rickye Heffner's blunt assessment when I rang her to discuss the research. Rickye epitomises the dogged experimenter vital to science. Imagine the painstaking training needed to get species as diverse as elephants, fruit bats and gerbils to provide reliable experimental data as to where they sense a sound is coming from. Sometimes it took nearly a year to get the data on a single species.

The poor localisation ability of horses was very surprising and at first Rickye assumed that there must have been a problem with her experiment. Imagine a horse at a water hole hearing a twig snap: surely it must be useful to locate the source of the sound aurally? One professor advised her, 'no one else is going to believe this unless you can demonstrate it three ways from Sunday'. Having rerun the tests with different horses and different protocols, she felt confident enough to publish. Reactions to the results were mixed, however; some were not convinced. The only way to win over these sceptics was not only to provide the evidence but to come up with an explanation for the results.

One night lying in bed and mulling it over, it occurred to Rickye that 'the whole point of having ears is to detect an animal and then tell our eyes where to look at it'. Maybe the selective pressure driving the evolution of sound localisation relates to the width of the visual field where the animal's vision is most acute? Horses have excellent horizontal vision over 180 degrees. Thus their ears do not need to give accurate information about location because their eyes can do that. All they need from hearing is the sensitivity to hear quiet sounds. Humans are different. Our best vision comes from the very narrow area served by a small depression in the retina, and this field of vision is only one to two degrees wide. We need good sound localisation to allow us to accurately orientate our eyes.

The most important graph from the Heffners' work shows results for about thirty species of mammals and indicates an impressive correlation between the accuracy with which an animal can localise a sound and the width of the field of best vision. At one end are humans, at the other animals like horses. I asked Rickye whether this graph had now won over the doubters. 'I guess I'd like to think so, but, maybe they all just died off,' was her wry answer. 'I was young at the time. You know that's how you win in the end, you outlive your enemies!'

As the Heffners' work shows, our exquisite hearing sensitivity arises to localise sound and so allow us to hunt and avoid predation. But what about the external ear, the pinna? What evolutionary pressures created such a distinctive shape? Again this is about localisation. Both ears hear identical signals for sounds emanating in front and behind because the head is symmetrical. Mixing up whether a well-camouflaged animal is in front or behind could be disastrous: you might run right into the claws of a predator. The pinna's asymmetric shape means that the sound from the front and the back is different and this helps to resolve front–back confusions.* As Rickye put it, the pinna is 'so boring, because it's just this flap of skin and cartilage sticking out there that people don't pay much attention

* The pinna and also the reflections from the shoulder are vital for detecting whether a sound is above you. Something very important to small rodents scurrying around on the ground trying to avoid large dinosaurs above them.

to. But it plays a big role in our ability to localise sounds.' As the human pinna is small, this localisation cue needs high-frequency sounds, however. This helps explain why we can hear beyond the bandwidth of speech.

Drawings of early mammals and their ancestors often include pinnae, but this is artistic licence: normally the outer ear does not fossilise. The earliest pinna fossil belonged to spinolestes, a mouse-like creature that lived in a swampy environment and whose diet probably consisted of small insects and animals it dug out with its powerful back legs. Found in Spain in 2015, this 125-million-year-old mammal was alive at the same time as the dinosaurs. The body was amazingly well preserved. Not only was one pinna found, but also spines reminiscent of a hedgehog, fur and hair follicles, as well as internal organs.[17]

As being able to hear is vital for localising prey and predators, the main building blocks of human hearing were in place millions of years ago, and a long time before speech could have emerged. Until recently, that would be the end of the story, but then scientists dreamt up an ingenious way to estimate the hearing acuity of early human species from fossils. Intriguingly, it turns out that hearing abilities have changed in an important bandwidth for speech. Was this a response to a newfound ability to speak? Or was it just a by-product of other selective pressures as humans evolved?

Rolf Quam, a palaeoanthropologist from Binghamton University, and his collaborators used CT scans to extract and estimate the dimensions of extinct hominin ears. They then applied a physics model to predict how sound waves would have moved the ancient ossicles and so infer the sensitivity of the hearing. They examined fossils from two early southern African hominins, *Paranthropus robustus* and *Australopithecus africanus*.[18] Both hominins lived in a woodland and savannah environment and had relatively small brains compared to modern man.* *Australopithecus africanus* lived a little over 3 million years ago, and the Taung Child skull from this species was the first pre-human ancestor ever to be discovered.[19] *Paranthropus robustus*'s name comes from its large lower jaw and molar teeth; it lived more recently, about 1.5 million years ago. Reconstructions of the faces of both hominins depict a combination of ape and human features, resembling characters from *Planet of the Apes*.

These early hominins had hearing bones that are somewhere between a modern human and a chimpanzee. The malleus is like that of a modern human, whereas the incus and stapes are more primitive, like those of a chimp. Also the ear canal has a different shape compared to both modern humans and chimps. These features probably gave the early hominins a bit more amplification in a bandwidth important for speech, about 1,500–3,000 Hz.[20] But these hominins are too ancient to have language, so this improvement in hearing in comparison with chimps must have occurred for other reasons. Quam has suggested that it might be about improving short-range communication on the savannah for simple vocalisations.

Studies have also examined more recent hominins.[21] *Homo heidelbergensis* reveals skeletal features that are closer to modern humans. The first human species to inhabit colder climates, it started about 700,000 years ago, and it is probably the last common ancestor of modern humans and Neanderthals.[22] European populations of *Homo heidelbergensis* then evolved into Neanderthals about 120,000 years ago, whereas a separate population in Africa evolved into *Homo sapiens* about 200,000 years ago.[23] The

---

* There has been a recent change in the taxonomy used to describe our evolution, leading to potential confusion. 'Hominin' refers to modern humans, extinct human species and all our immediate ancestors. 'Hominid' is a bigger group including the great apes.

hunt for the most recent common ancestor plays an important role in understanding evolution. In this case, the similarity in the hearing abilities of *Homo heidelbergensis* and *Homo sapiens* implies that Neanderthals were perfectly capable of hearing speech. That conclusion was confirmed by tests on Neanderthal ossicles. In 2016, Alexander Stoessel from the Max Planck Institute for Evolutionary Anthropology in Leipzig and his collaborators showed that despite Neanderthals and modern humans having slightly different hearing bones, both configurations would have given similar hearing abilities.[24] It seems that any adaptation of the middle ear in response to vocalisations was completed when *Homo heidelbergensis* appeared at least half a million years ago.[25] Therefore, speech evolved and took advantage of existing auditory abilities rather than the other way around.[26]

*

The evolution of spoken language is much more controversial than that of hearing. Today much of the controversy centres on the role of the Neanderthals living in Ice Age Europe who died out about 35,000 years ago.[27] *Homo sapiens* migrated from Africa to the rest of the world about 60,000 years ago. As language existed before *Homo sapiens* left Africa, this means modern humans were speaking when Neanderthals were still extant.[28] Neanderthals clearly had the ability to hear what we were saying – but did they join in our conversations?

On one side of the debate are those who argue that language arose relatively recently with the arrival of *Homo sapiens*, and that it was this linguistic ability that allowed us to outcompete other prehistoric humans.[29] On the other side are scientists who claim that Neanderthals were smarter than previously thought, had some language ability and interbred with our ancestors rather than being simply wiped out by them. Some would go even further, claiming that *Homo heidelbergensis*, the common ancestor of both species, was able to talk. If this is true, hominin language could have arisen hundreds of thousands of years ago. Thus the two rival theories date the start of language somewhere between 700,000 and 70,000 years ago – a time span of over half a million years! What is the evidence on both sides of the debate? Can science ever hope to resolve this argument?

Our basic speech mechanisms are not so very different from how other mammals vocalise. Consider voicing a simple vowel sound like 'e'. To do this, air is pushed up from the lungs and passes through the vocal folds (often referred to as the vocal cords), which are located in the larynx (the voice box). The vocal folds rapidly open and close cutting up the air flowing out of the lungs and so create a buzzing sound. The rate at which the vocal folds open and close determines the pitch of the speech. For instance, an adult female typically opens and closes her vocal folds about 200 times a second, producing a frequency of 200 Hz. (The frequency for males is lower, at about 110 Hz.)

The buzz from the vocal folds then goes into the vocal tract, the name of the air space including the top of the throat, mouth and nasal passages, and this then alters the sound. Like most sounds, the buzz from the vocal folds contains both the fundamental frequency and also harmonics that are multiples of this – 400 Hz, 600 Hz, 800 Hz, etc. These harmonics are vital to speech, because it is the relative strengths of these that are manipulated by the throat, tongue, mouth and nasal passages to get different vowel sounds. It is the deftness and speed with which we can alter the vocal tract that separates us from other primates. Our cognitive power allows humans to make incredibly rapid and complex changes to the vocal tract, coordinated with changes to breathing and the muscles supporting the vocal folds, to create fluent speech.

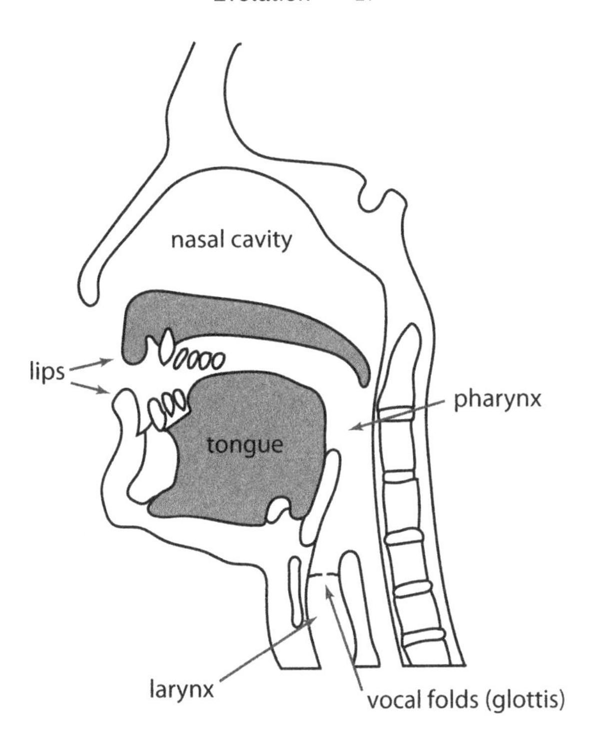

The vocal anatomy.

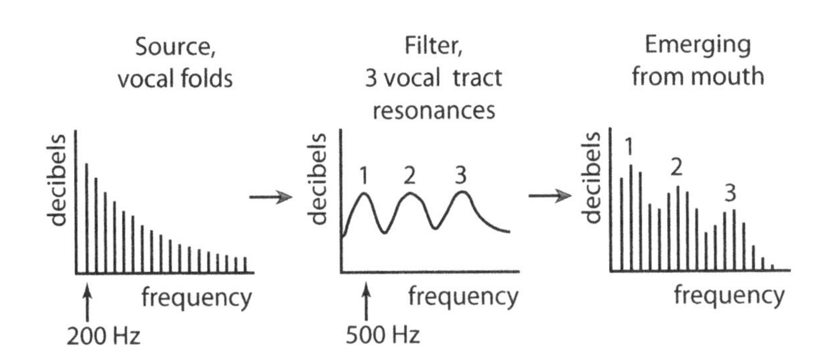

Creating a vowel sound with a frequency of 200 Hz and a first formant resonance of 500 Hz.[30]

The vocal tract is like the air column inside a trumpet: it has a set of frequencies at which the air inside vibrates loudly. These are resonant frequencies, and any harmonics from the vocal fold buzz that coincide with these are amplified. (Other harmonics fare less well and get suppressed.) The resonances of the vocal tract are called *formants*. Say 'hot – hat – hit' and you will notice how your mouth changes shape with the different vowels. The soft palate, tongue and lips, which are collectively known as the articulators, are shaping the vocal tract to get the formants right for each vowel.

It is perfectly possible to talk in a dull monotone and produce intelligible speech by just changing the formants through the articulators. As the pitch comes from the vocal folds, these can be made to oscillate in the same way for every word. This certainly worked for Clint Eastwood in *The Good, the Bad and the Ugly*, where his character Blondie has a breathy, husky, monotone voice.[32] As Eastwood demonstrates,

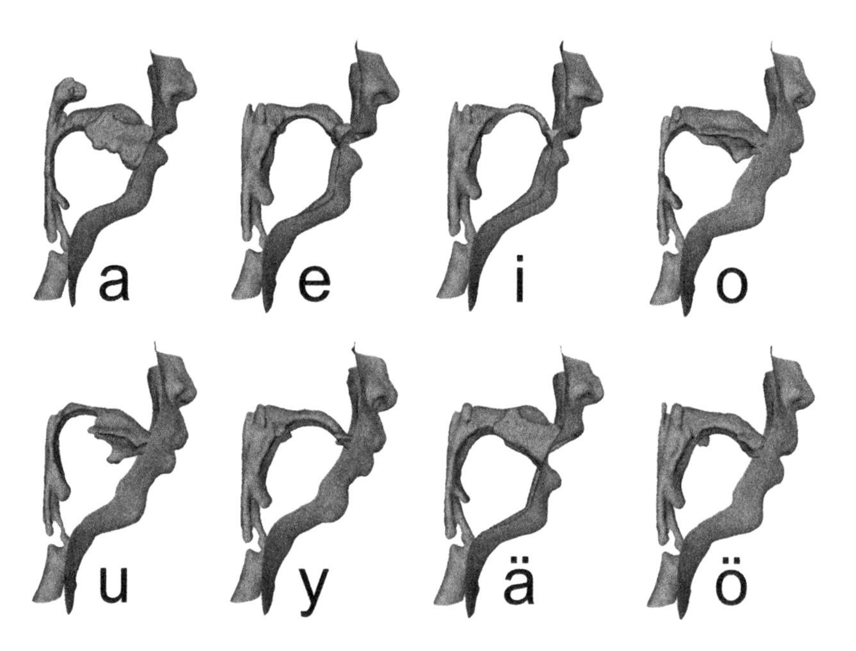

The vocal tract needs great flexibility to change shape for different vowels. Measured in an MRI scanner.[31]

the pitch of a voice is different from the formants that filter the sound and tell the listener what vowel is being said. Another good demonstration of this is the singing synthesiser heard in hit records such as ELO's 'Mr Blue Sky' and Daft Punk's 'Harder, Better, Faster, Stronger'. In this case, music production tricks are used so the vocal fold buzz gets replaced by musical notes, while the formants that allow us to hear and understand the lyrics are left untouched.*

To better understand the evolution of speech, it can be revealing to compare humans to other extant species. There are two important differences between how chimps and modern humans make sound. The larynx of modern man is much lower than a chimp's, and apes have air sacs alongside their throat. Many researchers have tried to pinpoint when the larynx descended, hoping that this is a unique marker for when speech evolved.

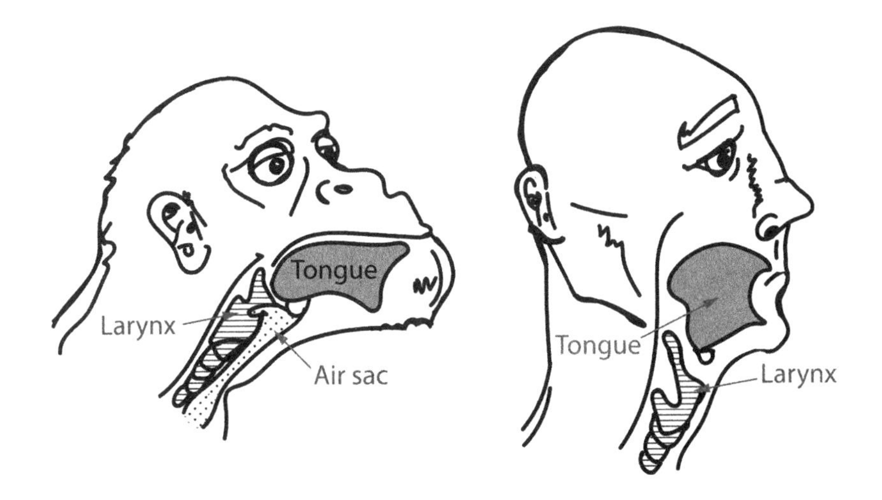

The larynx is higher in a chimp compared to a human, and a chimp also has an air sac.

* 'Mr Blue Sky' used a Vocoder. Unfortunately, Daft Punk are coy about revealing exactly which technology they use. In a later chapter I will examine the two most likely technologies.

In most mammals, the larynx is high enough to allow breathing through the nose while simultaneously swallowing. This is also essential for human babies who need to suckle and breathe at the same time. Between the ages of three months and four years, the human larynx descends to a lower position.[33] Men's larynxes then descend further during puberty.

This lower larynx is vital because of what it allows the tongue to do: without this we would not be able to make the vowel sounds in 'beet' and 'boot'. The lower position lets the rounded tongue have great mobility, allowing fast alterations to the upper throat and mouth that are vital for rapid and clear formant changes while we are speaking. With a lower larynx, the tongue root is pulled downwards, and this allows the pharyngeal cavity (the top of the throat) to be altered independently from changes to the oral cavity. Without this, our speech would be slower and imprecise.

In his book *The Biology and Evolution of Language*, Philip Lieberman describes a simple experiment that demonstrates the efficiency of speech in transmitting information. To do this you need a friend to help out. Get your friend to tap a pencil as fast as they can, while you estimate the rate of tapping by counting how many taps are done in five seconds. Your friend will be able to tap much faster than you can count, especially if they practise for a while. The fastest counting rate is about seven to nine taps per second. But when listening to speech we pick up changes at a rate of about twenty to thirty sounds per second, about three times faster. When saying a word like 'bat', we do not pronounce each letter separately ('b-a-t') because it would be too slow: instead we allow the sounds of the letters to run into each other, allowing information to be conveyed speedily.

The oral and pharyngeal tubes need to be about the same length to create fluid and distinct speaking. The horizontal distance from the lips to the back of the throat should be the same as the distance from the vocal folds to the soft palate (the back roof of the mouth).

Then the highly mobile tongue can change the cross-sectional area of the two tubes independently. Watch an MRI video of a cross-section through someone's head and at rest the tongue is a round blob. But as soon as the person starts talking, the tongue rapidly morphs into different shapes, dancing around back and forth and up and down changing the shape of the vocal tract. Try saying 'see' then 'ma' and note how the tongue position changes when making the vowel sounds. For 'see' the tongue moves up constricting the oral cavity, in 'ma' it drops and creates a wider tube. It is harder to sense how the vertical pharyngeal tube changes. For 'see' the tongue is moved forwards in the mouth and this opens the pharyngeal tube, but for 'ma' the tongue retracts to constrict the pharyngeal tube.

The vowel in 'see' is called a super vowel because it plays a crucial role in being able to understand different people speaking. Everyone's vocal tract is different, and this means the frequencies of the formants vary from person to person. If you say 'bit-bet', the differences in the vowel frequencies are clear and this helps a listener work out what word is being said. But the difference might not be clear between two different speakers. When a short person says 'bet' there is potential confusion with a tall person saying 'bit', because the differences in the vocal-tract lengths could render similar first formant frequencies. To avoid this confusion, the listener subconsciously estimates how long the vocal tract of the speaker is. When saying the vowel in 'see', the tongue is pulled up and as far forward as possible. Try saying 'see' normally and then inch the tongue forward a bit; the sound begins to buzz. This super vowel puts the tongue in its most extreme position: it can go no further while creating a clear sound, which allows the listener to estimate the length of the vocal tract of the talker and so calibrate their perception.

The relative lengths of the oral and pharyngeal tubes play a crucial role in talking fluently, so discovering these dimensions in our ancestors could be useful in understanding the evolution of speech.

But measuring these lengths for extinct hominins is problematic. The vocal apparatus is hung from ligaments and muscles attached to the base of the skull, and these do not fossilise. The hyoid bone (the U-shaped bone that anchors the tongue root) is the only structure that might be preserved, but controversy surrounds the interpretation of the data. Besides, it is not directly attached to other parts of the skeleton and so it is often missing from fossils. One 60,000-year-old Neanderthal fossil from Israel shows a hyoid bone that is the same shape as that of modern man.[34] Samples like this are incredibly rare and so the discovery created much excitement. But given that the shape of the hyoid bone can only ever be a poor indicator of vocal ability in hominins, even a large collection of hyoid fossils is unlikely to cast much light on speech evolution.* More useful is a fossil of *Homo heidelbergensis* that is at least 530,000 years old. It has seven cervical vertebrae in good condition and this allowed the Spanish palaeontologist Ignacio Martínez and his colleagues to examine the length of the throat and thereby estimate some dimensions for the vocal tract. They concluded that the dimensions were akin to a modern ten-year-old child.

Developmental studies of modern humans support the idea that the vocal anatomy of *Homo heidelbergensis* would have been capable of speech. As an infant grows up, the ratio between their horizontal and vertical tube lengths in the vocal tract changes from roughly 1:1½ at one month old, to the ideal ratio of 1:1 at about age nine.[35] A child's diction is not as good as an adult's, but they start talking long before the age of nine. This shows that even if early hominins had not got a fully descended larynx, their vocal apparatus would

---

* Furthermore, if the hyoid had changed shape during evolution to facilitate speaking, then why is a similar change not mimicked during human development? Evo-devo would anticipate a likely change in the hyoid shape in modern infants and teenage boys as the larynx descends, but it does not.

not have prevented them from talking. Still, their talk might have lacked the fluency of modern speech.

A recent groundbreaking study has gone further and demonstrated that a monkey's vocal anatomy is perfectly capable of creating words. So why do they not speak? The answer is that they lack the cognitive power to control the articulators. Tecumseh Fitch from the University of Vienna and collaborators produced simulations of what a macaque monkey's vocal tract could do if they had better muscle control.[36] They took X-rays of a monkey when it was doing things like cooing, smacking its lips and eating. With this information, they worked out what shape its vocal tract could make; and from that they calculated formant frequencies. The results showed that the monkey can form a wide range of formant frequencies and vowel sounds. To emphasise this point, Fitch put this information into a speech synthesiser to show what a monkey vocal tract could do if unconstrained by cognitive limits. I am not sure why the phrase 'will you marry me?' was chosen for the demo, but what emerges sounds like Gollum from the *Lord of the Rings* proposing. It is not as clear as when humans talk, but the experiment demonstrates that non-human primates have vocal tracts that could produce intelligible speech.

These studies seem to undermine the hypothesis that the descent of the larynx is a useful marker for speech evolution. But the idea cannot be simply dismissed without an alternative thesis for what drove the anatomical change. And whatever the evolutionary driver, it must have been strong enough to compensate for the risk generated by the fact that a lower larynx makes it more likely you will choke. Humans are not the only animals with a permanently lowered larynx, however: koalas and Mongolian gazelles have this feature too. Dogs and other animals also temporarily lower their larynx while calling.[37] While a dog normally has a raised larynx, when barking the vocal apparatus is rearranged to resemble more closely that of a human. Why?

When the larynx lowers, the vocal tract is lengthened and the formant frequencies in an animal's calls go down. This suggests a larger animal. The signalling of body size would be a strong selective pressure to encourage the larynx to lower either permanently or temporarily while calling.[38] How the human vocal anatomy changes during puberty also lends weight to these ideas. This second descent of the larynx, which occurs only in males, is not accompanied by improved speaking ability – as anyone who knows a teenage boy will testify. It probably occurs to give adult males a more booming voice, to exaggerate their body size and so help attract females.[39] (I will return to the subject of attractive voices in a later chapter.) It therefore seems that in hominins, the larynx probably lowered to attract a mate rather than to aid speech.

Gorillas, chimpanzees and bonobos have air sacs in their throats, something that humans lost at some point during evolution. These air sacs create an additional low frequency formant that radiates efficiently through the walls of the throat and creates the impression of a larger, more impressive animal.[40] The hyoid bone of apes has a small cup-shaped extension (bulla) that is thought to play a role in keeping the connection between the vocal tract and air sac open. As *Homo heidelbergensis* lacks the bulla on the hyoid, it seems to suggest that the air sacs disappeared from hominins between 3.3 million and 530,000 years ago. Maybe this was to aid intelligible speech because the ape's air sac alters how the vocal folds vibrate. Bart de Boer from the University of Amsterdam tested how the air sacs might have inhibited speech.[41] He created a set of Perspex tubes modelled on the shape of the vocal tract for particular vowels. Some had an air sac, others did not. Playing a typical vocal-fold buzz through the tubes allowed him to create renditions of different vowels. When played to listeners, it was found that the different vowel sounds were harder to distinguish from one another with an air sac. The implication of this is that air sacs might have been lost to help fluent talking. It seems that our vocal anatomy was good enough for some form of speech more than 500,000 if not millions of years ago. While science now provides a more complete story for the evolution of the vocal anatomy, the evidence can only go so far in helping us date the evolution of language.

*

If studies of the vocal and hearing anatomy have their limits, what other evidence could we look at? In *The Descent of Man*, Charles Darwin wrote that the study of language evolution should focus on improvements in cognition rather than changes to vocal and hearing anatomy: 'The fact of the higher apes not using their vocal organs for speech, no doubt depends on their intelligence not having been sufficiently advanced.'[42] He also noted that 'parrots are notorious imitators of any sound which they often hear', and so argued that articulation alone is insufficient to explain the complexity of language. Nearly 150 years have passed since these observations and today we know much more about the evolution of speech and hearing; yet Darwin's recommendation to focus on cognition is still sound advice.

Drawing on his empirical observations, Darwin argued for a sung protolanguage, an ancestor to modern speech that did not yet contain all the grammatical elements of a full language. This required an improvement in hominin mental abilities, Darwin believed, which then allowed the sung protolanguage to be refined as it was used in mate selection; finally more complex meanings were attached to the songs as intelligence improved further, partly driven by the evolution of language itself.

The idea that an increase in cognitive power came before the development of protolanguages is consistent with many modern theories. The *Homo sapiens* brain is about three to four times bigger than a chimp's, with the changes starting about 2 million years ago.[43] In the Natural History Museum in London there is a hominin family tree from the last 7 million years set out with sixteen casts of skulls. When I stared at the skulls from the front, it was the changes around the eye sockets that first jumped out to me, with many hominins having a much thicker brow ridge than *Homo sapiens*. Look at the skulls from the side, however, and the differences in cranial capacity become more obvious. The back of the skull is larger and more bulbous in more recent species of *Homo*. An analysis of cranial capacity has to be done with caution, however, because as *Homo* became larger, the brain naturally swelled to service the bigger body. Most intriguingly, Neanderthals have a cranial capacity about 10 per cent larger than *Homo sapiens*.[44] They were stockier and heavier, probably as an adaptation to survive the colder conditions in Europe, which would explain the bigger brain.

Darwin believed that the protolanguage had many similarities to birdsong and was used to attract mates, defend territory and express 'various emotions such as love, jealousy, and triumph'. Our ability to mimic sounds is important for the evolution of speech; it sets us apart from other primates who have a very limited ability to go beyond their species' vocal repertoire. Vocal imitation can be found in other species such as humming birds and bottlenose dolphins, however. The superb lyrebird is famed for its ability not only to imitate the songs of other birds, but sounds it hears in the Australian rainforest, including the click from tourists' cameras, chainsaws cutting down trees and the high-pitched screech of car alarms. Darwin believed that human language would have first started with analogous imitations of natural sounds, and the calls of humans and other animals. With enhanced cognitive ability, however, humans could attach more complex meanings to the sounds. As Darwin wrote, 'may not some unusually wise ape-like animal have imitated the growl of a beast of prey, and thus told his fellow-monkeys the nature of the expected danger? This would have been a first step in the formation of a language.'[45] As the quote implies, there is more to the development of language than sexual selection. In most animals the male is the more prodigious vocaliser during courtship, and yet in humans language is equally strong in males and females because of the survival advantage offered by being able to pass on knowledge.*

Given that vocal mimicry has evolved independently across diverse species, it is likely that hominins could imitate sounds as speech emerged. How might we have got from imitating the roar of a lion to the complex grammatical constructions of modern language? One possibility is that we started with vocalisations that conveyed very simple messages such as 'Danger: predator!'[46] Over time our ancestors' brains started to break down these vocalisations and began ascribing meaning to different parts of the sound; from this the nouns, verbs and adjectives and other linguistic syntax we use today gradually emerged.[47]

Lacking a time machine, how can science test whether this is what happened? Over the last couple of decades new methods have been devised with one being especially powerful: computer simulation. This allows researchers to explore what-if scenarios and test ideas for how syntax might emerge. One of the pioneers of this work is Simon Kirby from Edinburgh University, whose work built on groundbreaking studies conducted in the 1980s by his PhD supervisor Jim Hurford.

Simon's computer creates populations of talkative individuals called 'agents'. Like a real population, agents change over time, as some die and new ones are born. They also breed and pass on information to their offspring. As the experiment progresses, something miraculous happens, as what begins as random chatter between

---

* Some have used the vital role of singing in forming bonds between mother and baby as an argument that sexual selection played a secondary role in the evolution of language. There is more on motherese in Chapter 2.

agents coalesces into a simple language. The actual phrases the agents start with are just random strings of text, for instance 'kihemiwi'. Each of these phrases are attached to a meaning. For example, each text string might represent a particular geometric shape moving in a certain way, e.g. 'a square moving left to right'. As they go about their 'lives', each agent hears utterances from others; their task is to memorise them and their associated meaning, and repeat them later. Of course, if each agent can store the string of text and the meaning exactly in the computer's memory, then recall will be perfect and nothing interesting happens. But if the computer is given only the ability to memorise and recall phrases imperfectly, then the agents are forced to seek more efficient and robust ways of encoding meanings in the text. Over time, parts of the text in a phrase start to take on specific connotations, the equivalent of words, because this is much more efficient. Take the sample text 'kihemiwi'. The end of this, 'miwi', might take on the meaning 'square', and the start 'kihe' describing the motion 'moving left to right'.[48]

Simon explained to me how these groundbreaking simulations were initially met with derision, but gradually became more accepted because they were 'an antidote to poor intuition'. The problem with complicated systems is that it is hard to divine what kind of global behaviour might emerge. It is quite a leap of faith to suggest that language structure could just fall out of a set of simple rules for how agents pass on and memorise messages, but the computer model shows that this is possible. As you might imagine, many researchers were sceptical about the approach, but then a stroke of luck led to an experiment that brought widespread acceptance.

This new experiment, the first to show cultural transmission of language between people in the laboratory, was undertaken reluctantly. Simon was supervising Hannah Cornish, a student who was required to do a physical experiment to meet degree regulations. Cornish really wanted to run more computer tests but that was not an option. So she hit upon the idea to re-enact one of her computer simulations with real people in the laboratory. The experiment sounds a bit like an involved game of Chinese Whispers using nonsense text. 'We thought that won't work,' was Simon's blunt assessment, but to his surprise it was a success. The results matched the computer simulations with random strings of text morphing into a simple language. Other researchers could now grasp what was going on because it involved real people rather than obtuse computer code and algorithms. Nowadays, it is a well-established technique to use both computers and real people in this type of study.[49]

There are very plausible arguments that gesture was an important part of the early language. Maybe it was the first language, but talking is more versatile than signing as it allows communication at night and frees up the hands for other activities. It is also more efficient as it uses less energy. Given that early hominins had vocal abilities like other apes, selective pressures would have favoured a verbal protolanguage. Once this protolanguage had developed, a virtuous evolutionary spiral would form, with the neural connections to the vocal anatomy improving to allow more rapid speaking, while language allowed us to think long trains of thought that drove improved cognitive abilities. But much of the evolution of language would be cultural rather than biological. Once languages come into play, individuals can adapt their behaviour to improve their chances of surviving and reproducing through what they learn rather than via some genetic advantage. Biological natural selection gets swamped by what culture can achieve. Indeed, language has helped humans to largely bypass the slow processes of biological natural selection.

When did a simple protolanguage turn into comprehensive human language? Once again, only indirect information is available. Is it possible to date the transformation by looking for evidence of higher cognitive abilities that imply thinking using a modern language? Clues might be the use of pigments and personal ornaments implying symbolic artistic thought, or tools that require complex planning during construction, or ritual burial behaviour. The archaeological record was once thought to show a dramatic cultural revolution about 40,000 years ago, but this seems to stem from a bias of where most digging was done. In recent years, studies have been increasingly conducted outside the developed world and evidence of abstract art has been found from earlier times. Various African archaeological sites have yielded pierced shell beads from about 100,000 years ago, and South Africa's Blombos Cave includes geometrically engraved plaques from about 80,000 years ago.[50] It appears that the changes in the archaeological record 40,000 years ago were caused by modern humans spreading out of Africa rather than by a step change in cognitive abilities due to language emerging.

What is clear from the archaeological record is that Neanderthals had little art and displayed hardly any other symbolic behaviours. There are a few pieces of pigments and possibly some etchings, but that is it. In contrast, the modern humans living alongside the Neanderthals had musical instruments and painted beautiful murals. For many, this is strong evidence that modern humans had much more sophisticated language than Neanderthals.

In the past, it was generally assumed that only *Homo sapiens* was highly intelligent. In 1866, Ernst Haeckel produced an evolution tree that showed a species called *Homo stupidus* immediately before *Homo sapiens*. When the first nearly complete Neanderthal skeleton was unearthed, 'the Old Man of La Chapelle', it was reconstructed to look more like an ape than a modern human.[51] In *The Outline of History*, H. G. Wells suggested in 1920 that 'Unlike most savage conquerors, who take the women of the defeated side for their own and interbreed with them, it would seem that true men would have nothing to do with the Neanderthal race.' The reasons for *Homo sapiens* not mating with Neanderthals were, according to Wells, 'extreme hairiness, an ugliness, or a repulsive strangeness'.[52]

Such prejudices do not fit well with findings from modern genetics. DNA provides evidence that 1–3 per cent of the genomes of many people living today come from Neanderthals.[53] The ability to extract DNA from fossils also allows researchers to examine the ancestries of ancient hominins. A jawbone from a 40,000-year-old modern man discovered in Peştera cu Oase (Cave with Bones) in Romania had a genome that was 6–9 per cent Neanderthal. Large segments of chromosomes were Neanderthal, meaning that this person had a Neanderthal ancestor about four to six generations back in their family tree.

As genetics offers incontrovertible evidence of interbreeding between *Homo sapiens* and Neanderthals, would they have done so if they could not communicate with each other? Of course sex does not have to be consensual and children and females might have been stolen in raids. But if relations were more amicable, that would imply Neanderthals and modern humans were able to talk to each other. As more DNA is gathered, it might be possible to gauge whether the transfer of genetic material came mainly from male or female Neanderthals, or both. This might help shed some light on the social dynamics between modern man and Neanderthals.[54]

Recently, Neanderthals have undergone a radical makeover, even more dramatic than if they had gone on the reality TV programme *Ten Years Younger.* The Natural History Museum has produced a full-sized reconstruction of a Neanderthal that fits with the description provided by the professor of human evolutionary biology Daniel Lieberman: 'A well-coiffed Neanderthal in a suit and hat could pass unnoticed on the subway.'[55] Shorter and stockier than modern man, the museum's Neanderthal is portrayed with a thoughtful expression and a hipster beard.

The field of language evolution is very fractious, with quite a few researchers arguing about each other's theories with open contempt for their rival's viewpoints. As Simon Kirby put it, 'almost everyone is holding a position that is likely to be wrong'. With trepidation, let me add my own reading of the evidence to the mix.

Many experts now conclude that Neanderthals had some form of protolanguage. This probably means that *Homo heidelbergensis* also had a protolanguage because it is the last common ancestor between Neanderthals and *Homo sapiens*. Given evidence of the fine-tuning of the vocal anatomy, brain size and interbreeding between Neanderthals and *Homo sapiens*, some form of spoken protolanguage seems likely to have been in existence about half a million years ago. But something changed when *Homo sapiens* came into existence about 200,000 years ago. An improvement in cognitive ability, maybe triggered by some genetic change, led to a virtuous circle, whereby more sophisticated language allowed more complex thought, and this allowed modern humans to out-think and outcompete Neanderthals.

Is there any chance of knowing for certain which theory is right? Scientists have already shown themselves to be extremely resourceful in inventing new ways of exploring language evolution, so I am hopeful of more complete answers in the future. Genetics should help us to understand the role of biological evolution in the story. There was great excitement when the FOXP2 gene was discovered in 2001 because it plays an important role in speech articulation. (There will be more on this gene in a future chapter.) A variant of FOXP2 has even been found in DNA extracted from Neanderthal fossils.[56] But before genetics can shed light on language evolution, scientists need to unpick the complex relationship between our genes and our ability to speak. We also require a lot more hominin DNA. The quest for speech evolution might be hampered by the fact that the vocal anatomy does not fossilise, but we still need archaeologists to keep digging.

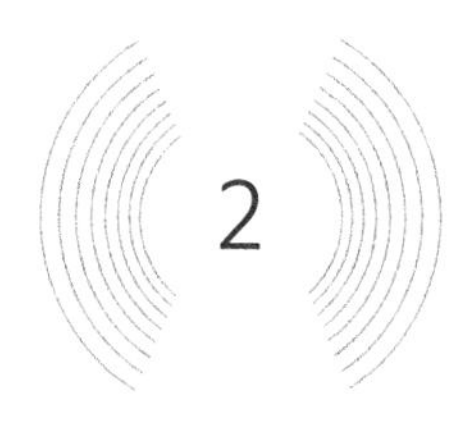

2

The Three Ages of the Voice

From the first cry at birth, through to the last words before passing away, the voice is life's accompanist. The average person speaks about 500 million words in a lifetime – and contrary to popular belief women utter a similar number of words to men.[1] I may have exceeded my quota already, however, as I was excessively talkative throughout my childhood. As my mum told me, 'Making your way against two older brothers, you didn't stop talking for quite some time.'

In his famous painting *The Three Ages of Man*, Titian depicts the three stages of life: infancy, adulthood and old age. Our voice goes through the same three phases, starting with childhood when language and speech begin to develop. Having become a radio presenter, it is funny to hear that when I was young I was assessed by a speech therapist. Apparently I did not pronounce the end of words. As my uncle Les once remarked, I spoke more and said less than anyone he knew. There was no underlying problem: I just wasn't bothering to articulate properly. Not everyone is so fortunate, as we shall find out when we come to explore this first age of the voice in more detail below.

The voice of the second age, adulthood, is most influenced by what happens during puberty as the body sexualises itself and the voice adapts to help attract a mate. Do you prefer the boom of Barry White or David Beckham's high-pitched chirring? Is breathy Marilyn Monroe more attractive than the sultry tones of Jessica Rabbit in *Who Framed Roger Rabbit*? While individual tastes are important, there are average trends that cut across the population, as we shall see. But what happens if normal puberty is prevented? Italian opera provides a grisly example as it turned castrated men into singing stars.

During the third stage, old age, the gradual deterioration of the body impacts speech. By examining some famous voices – such as those of Alistair Cooke, the Queen and Frank Sinatra – we shall see that although the voice is remarkably robust, even professionals eventually succumb to the effects of ageing.

You have a voice from the moment you are born. Before you enter the world, your lungs are deflated and filled with amniotic fluid. A foetus can hear in the womb from the third trimester, but it is impossible for it to talk. Doctors might not slap a baby's bottom to force a cry any more – more gentle measures are usually sufficient – but that crucial first shriek is noted by medics. It is an important sign of healthy respiration, forming part of the Apgar score used to assess the health of the new arrival minutes after birth. This is something I remember well, as one of my sons scored very low in the delivery room and was urgently whisked away for medical attention.*

At birth, the baby's voice is much more immature than its hearing. The larynx is poorly developed, so even if babies had the mental ability to talk, they would lack anatomical and neurological finesse to produce words precisely. The vocal tract is closer to that

* He is now a strapping adult.

of a chimpanzee than an adult human, with the larynx high up the throat. Between the ages of three months and four years, the larynx descends, which allows finer control over the tongue and better articulation of speech.

A study of nearly 9,000 babies found that they cry for about two hours a day for the first few weeks, dropping to just over an hour a day by week twelve.[2] At first the newborn cry is the simplest of vocalisations.[3] Each part of it involves the pitch and loudness first rising and then falling. The image below shows three individual examples on the left. The top trace shows a common representation of speech as variations in pressure over time. For vocalisations, the lower image is more useful because it reveals the variation in different parts of the sound. The harmonics in the cry show up as parallel dark lines that rise and fall in frequency, giving the audible change in pitch. The newborn has little control over the vocal folds, and so the sound is mostly determined by how hard the air is being pushed out of the lungs. The motion of the vocal folds is governed by the Bernoulli effect. Named after the eighteenth-century Swiss mathematician Daniel Bernoulli, it examines the interplay between the air speed and pressure. As air leaves the lungs, it has to speed up to get through the small gap in the glottis. As predicted by Bernoulli, this airflow then causes a drop in pressure that enables the vocal folds to shut. But then the air from the lungs forces the vocal folds open, followed by the Bernoulli effect closing them again, and so on. In the middle of the cry a baby is usually pushing air out more forcefully, so the vocal folds open and shut more rapidly, resulting in the pitch rising along with the loudness.

Surprisingly, very early cries already have some individuality. One study has demonstrated that they are influenced by what the foetus hears in the womb. Although a baby has little control over its vocal folds because of its immature nervous system, it can change

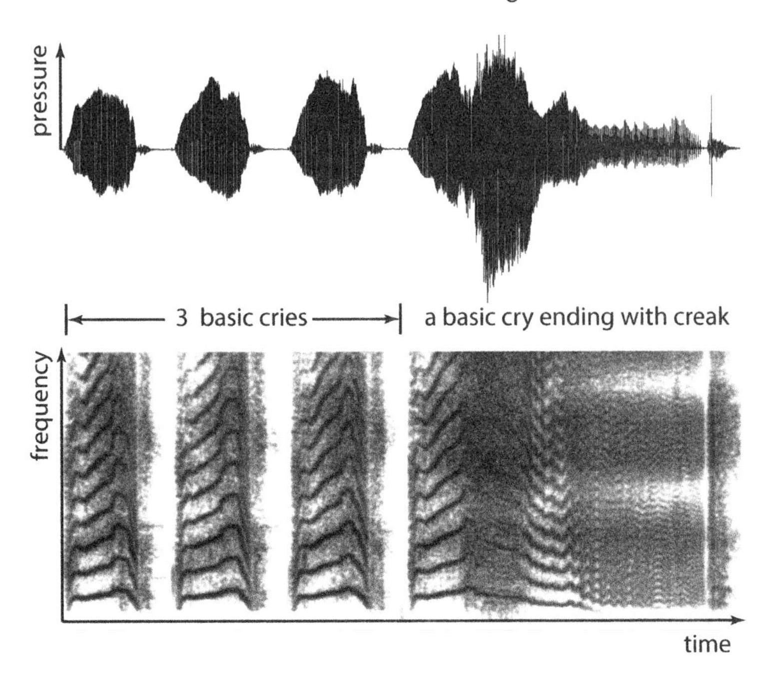

The acoustic signature of three basic newborn cries followed by a more complex cry that finishes with a creaky sound where the clear harmonics disappear.

how it breathes to alter how quickly both pitch and loudness first rise and then fall. Thus it can crudely alter the cry's intonation. The study was conducted by Birgit Mampe from the University of Würzburg in Germany and collaborators. They compared the cries from thirty newborn babies from France and Germany.[4] The French babies took their time reaching peak pitch and loudness, whereas the German neonates got there quicker; they therefore spent longer on the second phase of the cry where the pitch and loudness gradually decreases. The researchers showed that these intonations mimic the two native languages. In French, the pitch generally rises over a sentence until the last utterance when there is a drop; whereas in German, there is more of a general fall in pitch and loudness over a sentence. Each newborn had already started to pick up the intonations of their mother's voice by listening in the womb, ready to imitate them after birth. This is the first sign of an infant mimicking their mother's behaviour to foster bonding. After all, crying is important for survival because it enables the helpless neonate to co-opt support from adults. As any parent knows, it is very hard to ignore the emotive wail of their child.

We know that by week twenty-four of pregnancy, foetuses are able to hear and respond to sounds. A few weeks later, the foetus can hear over an octave, about a tenth of the frequency range of a young adult. The soundscape is dominated by the mother's bodily noises, such as the rumble of her voice, squelching and gurgling from the stomach and intestines, and the rhythmic thud of her heartbeat. All these sounds are muffled as they pass through the mother's body and amniotic fluid before they reach the foetus's ear – it must be a bit like trying to hear underwater in the bath. For a BBC programme I once interviewed a company that produced CDs which featured a simulation of sound inside the womb mixed with ambient music. They claimed these helped a child to sleep. I am not sure the CDs have been scientifically tested, but studies have shown that womb sounds can reduce heart rates and soothe premature babies in neonatal intensive care.[5] There are also hints that playing a muffled version of the mother's voice in the incubators could help the development of the auditory cortex of the preterm babies. As the study into neonate crying shows, foetuses pick up aspects of the rhythm and intonation of their mother's speech, so it makes sense that premature babies would gain from hearing voices as well.

What about babies that go to full term? You can buy loudspeakers that strap to a mother's stomach or are placed in the maternal vagina, but there is no good evidence that a foetus benefits from sounds played through these.[6] Quite the opposite: I would be concerned that some devices could pose a risk to hearing because we don't know how vulnerable foetus ears are to damage. Will playing music to a foetus boost or hinder neurological development? There is no scientific study to answer this question, so I believe it is best to let the foetus experience an old-fashioned maternal soundscape without augmentation. Incidentally, as the mother's voice is heard most often and most clearly, this explains why a newborn recognises her voice in preference to others, including the father's.

Even though it was some years ago, I clearly remember the parental challenge of trying to work out what my sons' cries meant, in a desperate effort to find out what would make them stop. A basic cry has a pitch between roughly 250–450 Hz, frequencies that are at the lower end of a violin's range. But sometimes a baby will suddenly switch to a much higher frequency – 1,000–2,000 Hz – producing a high-pitched squeal, a frequency towards the top of a violin's range. It has been compared to an adult singer switching from normal singing to falsetto.[7] A baby also has a third type of cry, which has a creaky quality. This is thought to be created when the vocal folds no longer open and close in a controlled, rhythmic manner.[8]

Over the first few months, the baby expands its repertoire by concatenating different cries and adding rhythmic variations. As the nervous system matures there will be less warbling as control over vocal muscles improves. The cries can then convey more complex information, showing whether the baby is in discomfort or hungry. Nowadays there are apps that claim to analyse the sound and deduce the cause of a baby's distress. Infant Cries Translator, for example, claims to have a 92 per cent success rate in identifying the causes of cries for two-week-old babies, differentiating between hungry, sleepy, stressed, annoyed or bored.[9] However, there is no independent scientific study of the effectiveness of these apps, and even the makers claim it is only useful for the first six months or so. I would be surprised if these apps did any better than a parent's own ears and discernment. And parents need to gain confidence that they can respond correctly to their baby's cries.

It is easy to dismiss crying as a simple distress signal, but the next time a baby throws a hissy fit, you should grudgingly admire the acoustic signature. A long tantrum is made up of a sequence of simple individual cries and there is a melodic line to it. Such intonation is a universal feature of language because it readily conveys emotion. When a baby puts an intentional pause in a cry by momentarily closing the glottis, it shows an ability to break up sound into chunks, which is a crucial skill needed later when the child starts uttering words.

While crying is the first verbal communication, fortunately babies soon start making cooing and babbling sounds. These are beautiful noises that are vital to the development of language, as an infant tries to copy what they hear. To speak requires the ability to hear and decode sound, as well as a nervous system capable of orchestrating the subtle movements in the hundred or so muscles controlling respiration, the vocal folds in the larynx and the vocal tract. Naturally, hearing needs to develop in advance of talking; doing it the other way around would be like trying to master a musical instrument before you had learnt to appreciate music. But an infant's immature attempts at talking also help it learn to decode what others are saying. The sounds of speech are limited by our vocal anatomy. A simple illustration of this is a tongue twister like 'she sells seashells on the seashore', which shows that there is a limit to how fast the vocal tract can change. Through talking, therefore, a listener learns about the limitation of their vocal anatomy, and this in turn helps work out how someone else's speech unfolds.

It is impossible to untangle whether early abilities to hear and decode speech are 'hard-wired' into the brain or come from experience. At birth the neonate can already perceive 800 or so phonemes. These are the building blocks of words and snippets of sound like 'a', 'wh' and 'ng'.[10] Given that hearing begins in the womb, it is difficult to separate what might be some predetermined processing structures in the brain, and what are neurological pathways that developed prior to birth from hearing the maternal voice. What has been shown, however, is that at birth we have a preference for sounds that are like speech.

Two Canadian researchers, Athena Vouloumanos and Janet Werker, studied twenty-two neonates who were one to four days old.[11] They played them nonsense words and tonal sounds reminiscent of the speech from the children's programme *The Clangers*.[12] They noted the reaction of the babies using a dummy hooked up to a computer that monitored the rate of sucking. The experiment was set up so that a sound was played every time there was a strong suck. The babies soon learned that they were rewarded with a sound when they used the pacifier. If this resulted in an interesting sound, then the baby would suck more to hear what comes next. If the sounds were repetitious, however, the babies got bored and stopped using the dummy so enthusiastically. The study showed that when the neonates were presented with speech, they sucked stronger for longer than they did for tonal sounds.[13] They clearly found the speech more interesting. A different study showed that up until the age of three months, babies also prefer certain monkey calls over non-speech sounds. This might be because the neonates are primed in the womb to pick up simple features of speech from the mother's muffled voice that happen to be also present in primate calls. It is only when speech processing in the brain gets more sophisticated a few months after birth, that the baby can distinguish the monkey calls from human speech.

The cooing that emerges in the early months is intuitively interpreted by parents as the baby being playful and friendly. When

I've interviewed people about their favourite sounds, baby cooing is often mentioned. Studies have shown that the more a parent responds to these sounds, the more the quantity and quality of the babbling increases. Ideally, therefore, there should be a duet between infant and parent to drive the development of the child's voice and language, social and cognitive skills. Unsurprisingly, infants soon learn to use their eye-gaze and babbling to get attention from others to help their learning. They even change their babbling to mimic the listener, with one study reporting babies using a lower frequency in the presence of the father compared to the mother.[14] For those who like to parent and multitask with their smartphone, beware: from four months old, infants have learnt when caregivers aren't paying attention! The infant is subtly manipulating other people's behaviour because it needs the right sort of information, such as the parent pointing at an object and naming it, to learn language.

This need for interaction was demonstrated in a study by Patricia Kuhl at the University of Washington who studied nine-month-old American infants.[15] Some of the children were made to learn from a live teacher, while others watched a video or listened to an audio recording of the same educator. Stories were read to the infants in a language they had not encountered before – Kuhl used Mandarin. This was done so she could distinguish what was heard and learnt during the scientific experiments from what the children picked up in their everyday experiences away from the lab. Kuhl monitored how well the babies remembered the phonemes for Mandarin after a month of listening to stories. She found the infants were interacting better with the storyteller in the room; both audio or video recording did not have the same effect. Could a robot do better than a video? The answer is probably yes, as social robots have been shown to be useful in supporting language learning. Hae Won Park, a postdoctoral researcher from the Massachusetts Institute of

Technology (MIT), works on Tega, a robot that looks like a Furby with eyes beaming out from a smartphone screen and actuators inside so it can bob up and down. In one test children had to tell a story to the robot. If Tega responded by leaning forward, nodding and smiling at the right moments, mimicking an attentive friend, the children told stories with more complex and longer narratives.[16]

My sons like to joke that my wife and I have no recollection of their important milestones such as their first words. My excuse is that we were parenting twins and were doing well to survive. While the first word is an important marker in a child's development, actually some of the most remarkable learning happens well before then. There is a rapid acceleration in the number of words infants can recognise as they home in on the forty or so phonemes needed for their mother tongue. The first year of life is a time of astonishing linguistic development even before any words are uttered.

An extraordinary experiment was started in 2005, when Deb Roy, a researcher with the MIT Media Lab, set out to record every waking moment of his son. He wanted to capture the emergence of language. Every corner of Roy's house was monitored by cameras and microphones, and most things heard or said by his son from nine to twenty-four months were recorded and transcribed, making a total of 8 million words.[17] This covered the period from the first word – 'Mama' – to the point where the infant was using combinations of words consistently. One dramatic piece of time-lapsed audio shows how the word 'water' emerged over a six-month period. At twelve months, Roy's son was using 'gaga' for water, but then that gradually changed. This sequence lists roughly two words per week:

> Gaga guga guga guga guga guga guga gega gugu guga guga guga wawa guga guga gugu wawa gaoo gaou yeya gogo wawa gaga gaga guga guga gaga wawat gugu gaga guga guwat gaga

> woda water gaga guga guga waki wooki wa chew wakri w doz
> vu cherk waa wa chew water[18]

At the end of the experiment, the two-year-old boy had produced nearly 700 unique words. Because every room in the house was monitored, it was possible to do a detailed analysis of when and where these emerged. Unsurprisingly, linguistically simple words emerged first, with 'fish' being learnt before 'breakfast', as did words that were uttered more frequently by those looking after the child. Also, words tied to a location or a particularly activity were learnt quicker, so 'bath' was learnt at eleven months, whereas 'head' emerged at twenty months.

Roy could also examine how he and others talked to his son – something that has been intensively studied by many others. Motherese is the sing-song voice people employ when talking to a baby, but the name is misleading because everyone does it: I remember using it on my children.* We often use the same exaggeration when talking to pets and (more embarrassingly) foreigners. In motherese, the melodic lines of the speech are exaggerated, the range of pitch widening as more high pitches are being used. This overstated intonation makes it easier for the infant to understand the emotional intent of what is being said.

In the first three months, motherese is dominated by melodic lines that gradually drop in pitch to soothe and comfort the infant, whereas later more complex trajectories get introduced. A rising pitch at the end of the phrase might be used as a hook to gain attention, whereas approval and encouragement tend to have a rising and then falling melodic line. Nairán Ramírez-Esparza from the University of Connecticut used audio recorders to monitor how adults talked to infants between the ages of eleven to fourteen months, later

* In the literature this is often called 'child-directed speech'.

measuring the infants' speaking abilities at two years old. Children of parents who used the most motherese had learned two and a half times more words than the infants of carers who used it the least.[19]

There is more to motherese than exaggerated pitch contours. Compared to normal speech, it tends to have fewer utterances, be louder and slower, with phonemes and pauses being lengthened. The words used are also simpler, with repetitions, and 'babyish' endings to words, such as doggie, horsie, etc. One of the most important listening skills a young infant needs to learn is how to break up speech and identify where words and syllables start and finish. If carers simplify, slowing and exaggerating the rhythm of their speech, it helps children to learn how to segment the stream of sounds into short snippets that can then be fully analysed by the brain. Motherese is complemented by motionese, a set of simplified, enthusiastic, repetitive and exaggerated gestures. This is often used by parents when demonstrating how to use toys such as stacking cups. If carefully synchronised, motionese and motherese can aid learning of speech.

What happens if an infant does not hear others speak? Over many years a debate has raged as to whether there is a critical period for language acquisition – the idea being that without exposure to language before puberty, acquisition would be permanently inhibited and no amount of intensive teaching afterwards could correct the problem. At one time it was thought that studying feral children who had suffered extreme language deprivation was a way of gaining insight into the topic. One tragic case involved a girl referred to as Genie in her case notes. In 1970 she was rescued from incarceration in Los Angeles.[20] She was thirteen years old, and had been locked in a small room, often tied to a chair and had heard virtually no words. Despite intensive speech therapy after being rescued, Genie could learn only rudimentary communication and lacked the ability to use grammar to string sentences together. But using such cases as evidence of a critical period is now thought to be problematic because of confounding factors. Genie's lack of language exposure was not her only problem: she had also been horrendously abused. Also, what if Genie had a pre-existing problem with learning before the abuse began? It is impossible to be certain that the lack of hearing speech was the only factor that inhibited her talking.[21]

There are infants with loving parents who have delayed access to language. Congenitally deaf children can be late in acquiring sign language because hearing parents first have to learn to sign before they can pass on the skill. In these cases, the signing of the children may never achieve the fluency or be as grammatically correct as those who were exposed to sign language earlier. It is somewhat similar to trying to learn a second language. In non-native speakers of English, grammar and vocabulary are worse if they started to learn the language after the age of seven than if they were exposed to English earlier in life. This also applies to comprehension. Most adult Japanese speakers struggle to perceive the difference between 'r' and 'l' in English, because the differentiation is irrelevant in Japanese. In contrast, six-month-old Japanese babies have no such problem.[22] It appears that we are born with the ability to work with any of the world's languages, but from about six months old, the brain starts to specialise in the speech sounds it hears every day. The infant needs to focus on the forty or so phonemes needed for a particular language and it does so by concentrating on the speech sounds it most often hears.

One of the largest studies in this area was done by Kenji Hakuta and collaborators.[23] They examined English proficiency of 2.3 million immigrants to America, who were native Spanish- or Chinese-speakers. The data was drawn from the 1990 US census, which included a question on language proficiency. The study showed that early exposure improved second-language fluency. Is this evidence of a critical period for language development?[24] Maybe not, as adult brains still have the ability to adapt because they have neural plasticity. However, learning is limited by our memories and our existing neural structures. Certain brain areas become optimised for a native language, and although there is potential for changing the brain, this flexibility reduces over time as neural networks get devoted to other tasks.

This is brought home by the sad cases of people who suffer brain damage in the left hemisphere. For most people the left hemisphere becomes specialised for linguistics at a very early age, dealing with many details of language.[25] The right hemisphere, on the other hand, usually deals with the intonation and stresses in speech, and how a discourse unfurls. If there is damage to the left hemisphere, say a focal lesion caused by a stroke, a tumour or trauma, then the brain has to reorganise itself. What happens varies between individuals, but infants often exploit neural plasticity so their right hemisphere takes over speech specialisation normally dealt with on the left. After time, they can speak reasonably well. For adults, however, such gross reorganisation is impossible, and hence a variety of speech impediments may result – the sort of effects that are all too common after a stroke.

Our understanding that the brain is divided into areas with very discrete tasks goes back many centuries. But in recent years neuroscience has moved away from the idea that functions are immutably tied to specific areas of the brain. The case of the man known as 'FV' illustrates this. If you feel your skull, just above and in front of the ear you will find a slight indentation. Behind the indentation on the left side is the inferior frontal gyrus that is important for your ability to talk. This brain region includes Broca's area, which is normally vital for language comprehension and speech production. Yet when a tumour was removed from FV's Broca's area, doctors were surprised how little impaired his language was.[26] It seems that the tumour growth was slow enough for his brain to adapt and use other regions for language.

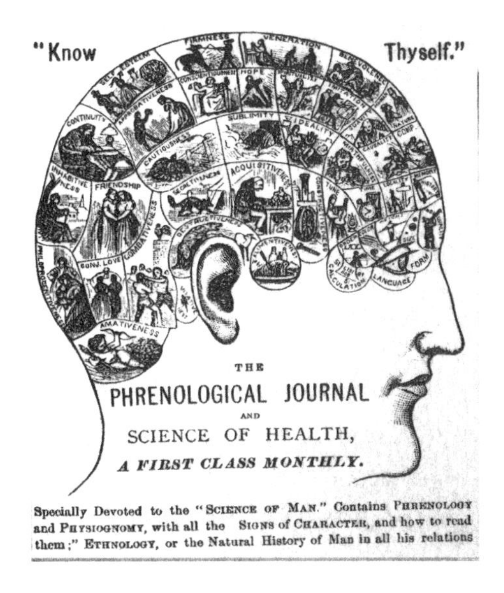

An out-of-date view of the brain divided into discrete regions each dealing with a different trait (date unknown but probably nineteenth century).

A modern view of the brain showing the emphasis on connectivity. (Courtesy of the Laboratory of Neuro Imaging and Martinos Center for Biomedical Imaging, Consortium of the Human Connectome Project – www.humanconnectomeproject.org.)

*

We make snap judgements about people based on the way they talk and this can be unfair on those who struggle with speech. Speaking differently from others, in particular, can create awkwardness in social situations: you're likely to be bullied at school and may develop psychological problems. Stammering is often portrayed as a purely psychological problem, something that could be fixed by therapy. But this is wrong. The underlying cause is now thought to be a neurological development disorder that often has some genetic component. Indeed, stammering, sometimes referred to as stuttering, provides an insight into how the brain deals with speech. It is relatively common in childhood, with one in twenty two-to-four-year-olds stammering at some point. This occurs during a phase in language development when word acquisition is extremely rapid, with the child learning about four new words a day. Fortunately, most people grow out of stammering so often the best thing a parent can do is not make a fuss about it. But if there are reasons to believe that it might persist, maybe because adult family members stammer too, then early intervention can improve outcomes. By adulthood, about one in a hundred people suffer from the condition.* To find out more about it, in 2016 I went along to the British Stammering Association conference at Manchester University.

One way to get an insight into the ailment is to use brain imaging to compare people with and without stammers. One study showed that in sufferers, the right hemisphere is more active during speech, maybe because it is trying to carry out tasks that are normally dealt with by the left hemisphere.[27] Further insight can be gained by

* Some people think this is a slight overestimation, but it is certainly the right order of magnitude, especially for males, who more frequently have the condition.

experiments where the fluency of speech is changed. A surprising feature of stammering is that there are 'tricks' to increase fluency temporarily. The effect is dramatic, as many videos on the Internet demonstrate. One technique features in the film *The King's Speech*, the movie that tells the story of how King George VI struggled with public speaking. In one scene, the speech therapist Lionel Logue blasts Mozart's *The Marriage of Figaro* over headphones as the king attempts to recite Shakespeare. The music prevents the king hearing himself talk and much to his surprise, he was able to recite the soliloquy 'To be or not to be' fluently.

Sophie Meekings is a PhD student from University College London. She has experimented with another technique called 'choral speech', where someone reads a text out loud in unison with others. This allows Meekings to change talking fluency and monitor what happens in the brain using fMRI. An fMRI scanner measures the amount of blood oxygenation in different parts of the brain using powerful magnets. When neural activity increases in an area, more blood flows there to replenish supplies, and so the oxygenation increases. This allows the scanner to identify which brain regions are most active while a person is speaking.[28]

Speech involves a precise, coordinated interplay of many different regions of the brain – its auditory, motor, cognition and emotion parts. When someone stammers, there are delays and difficulties in moving the speech forward. The problems arise because of imperfect connections between the brain regions dealing with speech planning and those controlling the vocal anatomy. Scans show that one brain area that is different for someone who stammers is the ventral premotor cortex, located roughly behind the eyebrows, about a third of the way into the brain. It is an area important for understanding, planning and executing actions, such as moving your hands, feet and mouth. But reduced activity in this brain region also implies difficulties in synchronising the planning and execution of speech – the person might be talking before they have fully planned what they want to say, for example. Alongside reduced neural activity there are also some structural differences in the brain of a person who stammers. There might be poorer connections between the ventral premotor cortex and the parts of the brain that process what is heard, for example. But focusing on this one part of the cortex is a simplification of the real condition: one study identifies sixty areas of the brain that can differ between people with and without stammers.[29]

Sophie Meekings has tested whether auditory feedback contributes to the condition. The hypothesis is that people who stammer are paying too much attention to their own speech, with the aural feedback that naturally arrives just after something has been said getting in the way of fluently producing the next word. This hypothesis seems to find confirmation in the scene from *The King's Speech* where loud music disrupts the king's ability to self-monitor his Shakespeare soliloquy. Another way of disrupting auditory feedback is to relay a slightly delayed version of a person's speech through headphones. For someone without a stammer, this usually makes them trip over their words and grind to a halt. But paradoxically, for someone with a stammer it can make them strikingly more fluent. Still, while these devices can be very effective for a few months, they are not a long-term solution because the brain eventually adapts and the feedback problem re-emerges.

In the film, King George objected to the speech therapist's approach, but he should have been thankful that some ancient treatments for the condition were not tried on him. The Greek statesman Demosthenes was supposed to have overcome his difficulties by speaking with pebbles in his mouth and reciting poems while short of breath. Given what we now know about the neuroscience behind the condition, the treatment is unlikely to have cured it. To find out about modern, more humane and effective treatments, I talked to the speech expert Christella Antoni. I first came across Christella when I saw her perform some amazing vocal impressions in a BBC TV documentary. She sang like Katie Melua, Barbra Streisand and Ella Fitzgerald and explained how she expertly shaped her vocal anatomy to create these different voices. Christella is a speech therapist who has worked with many people who stammer. She explained to me that there is no magic cure to remove a stammer; instead she advocates an approach that means it no longer dominates someone's life.

If you have a stammer, a common natural reaction is to try and avoid words that are problematic to say. If saying the word 'difficult' is a problem, you might substitute the word 'hard'. But over time, as more words have to be avoided, this evasion strategy becomes harder to maintain because there will always be words, like the person's name, that will need to be articulated; the author Lewis Carroll is supposed to have struggled to say his real surname Dodgson. One part of helping with the condition is to reduce such avoidance strategies. Christella tries to get people to accept that stammering will inevitably occur. This reduces stress levels and so decreases the chances of it happening. She described the case of an eminent researcher who had to give high-powered presentations and was understandably worried about his speech breaking down. Christella got him to mention his condition at the start of the presentation, to take away some of the stress and so reduce dysfluencies.

Many of the people at the conference I attended supported this approach. As an outsider who had never before thought deeply about the condition, it was insightful to encounter people who despite their stammer just got on with speaking. Initially I found myself willing them to finish words, but after a while I learnt to relax and wait for them to get to the end of their thought. A discussion with Patrick Campbell, a trustee of the British Stammering Association, gave me a new perspective. Patrick told me about his life including being 'therapied to the hilt' during childhood. He struggled in his first year of medical school. 'I tried not to speak,' he said with a chuckle, 'and you can't be a doctor who doesn't speak!' Now he is a successful junior doctor open about his speech which no longer causes him any problems.

In fact, Patrick questioned whether stammering should be seen as a 'defect' to be cured. He complained about the standard media trope exemplified by the TV programme *Educating Yorkshire*. In one scene the stammering student Musharaf Asghar used the trick of playing music over headphones, like in *The King's Speech*, to deliver a speech fluently. As the *Guardian* reviewer described the response: 'His friends and teachers sobbed. Viewers collapsed into a soggy pool of joyful tears. It was a beautiful triumph, a moment of pure elation, and it was one of the defining moments of television in 2013.' But Patrick was uncomfortable with the triumphalism. As he explained, 'The audience weren't prepared to see him just stammer through his talk on stage. That dodged the problem rather than solved it.'

Patrick is an advocate of what he terms 'dysfluency pride'. As he wrote in a blog post:

> What if, rather than hiding our stammer to appease society's demands, we fight for our right to stammer? Dysfluency pride is looking for individuals to question the opinion of society and take a much more empowering view of their speech. Already many forms of therapy aim for clients to accept stammering but dysfluency pride wants us to flourish with it. It wants us to stand up and question the current fluent values of society, to stammer loudly and proudly, to show society what we sound like.[30]

At the end of our conversation I asked Patrick whether he would want me to remove his stammer if I could wave a magic wand. The most telling aspect of his answer was his initial silence. He thought for an age before saying, 'I think I'd rather stammer, but that is on a knife edge.' After all, for better or worse, stammering had shaped his personality. Given that the condition is a fact of life and there is currently no magic cure that can give complete fluency, I agree that social attitudes need to change.

Stammering illustrates just how complex the act of speaking is. As talking fluently comes to most of us naturally, it is easy to forget how much brain power this takes and how miraculous the development of our ability to speak is. As stammering shows, it does not take very much to upset the vocal development and to change someone's voice. Given that this often leads to bullying at school, being inhibited from talking on the telephone, or joining in at parties, stammering does not just change a voice, it fundamentally alters people's lives.

How important are our genes to our voice? Genetics certainly plays a role in stammering. Lewis Carroll referred to his stammer as his 'hesitation'.[31] Carroll's parents were first cousins, and nearly all of their eleven children struggled with stammering both as children and adults.[32] Today it is estimated that between 30–80 per cent of stammering cases have a genetic component. So far, four gene mutations that might be the cause of missing neural connections have been identified.

Stammering is just one condition that can shed light on the importance of genetics to speech. The difficulties faced by some of the 'KE family' led to overblown media claims that science had found 'the grammar gene' in 2001. This British family attracted interest because of their inherited language defects. Roughly half the family had profound speech and language problems arising from difficulty in achieving the necessary fine-coordinated control of mouth, tongue, lips and soft palate. A mutation in a specific gene, FOXP2 on chromosome 7, seemed to be the cause. Current thinking is that FOXP2 is important for neural plasticity in regions vital for language development, but calling it the 'grammar gene' is a gross oversimplification. The connections between genetics and the complexities of speech remain difficult for scientists to untangle.

The importance of genes to the voice is unsurprising, given that DNA is the starting blueprint for human development. My genes, along with those of my wife, have influenced the size and shape of the vocal anatomy of our children, for example. We are a tall family, meaning that we probably all have larger than average vocal tracts, which means my sons are likely to have lower formant frequencies. But there are environmental factors as well. We bequeathed DNA but also spent much time chatting with our children. Brains adapt to what they hear, so parents' everyday speech influences their children's voices.[33] Still, there is a limit to such parental influence, which explains why my children have a 'Manchestahhh' accent, whereas my voice portrays my southern English roots, drawing out my vowel sounds in words such as 'baahth'. (We will learn more about accents in Chapter 4.) As everyone has an ability to change the size and shape of their vocal tract, genetics is only dominant when it comes to gender. And it is at puberty that the voices of males and females diverge. This is the time when we develop the voice that carries us through the second vocal age of adulthood.

*

During puberty testosterone triggers a thickening and lengthening of the male vocal folds. This causes the voice pitch to drop typically by an octave, an interval like the big jump at the beginning of 'Somewhere Over the Rainbow' – only the other way: descending from 'where' to 'some'. At the same time, an enlargement of the vocal tract also lowers the formant frequencies, further changing the quality of the male voice. What is less commonly discussed, however, is how the female voice changes. Typically, a teenage girl's vocal folds lengthen by about a third and thicken, and this helps drop the fundamental frequency of the voice about three semitones by adulthood – that is, the interval between the first two notes of 'Swing Low, Sweet Chariot'. But pitch is not the only characteristic to change. For about a third of young women, the vocal folds do not always completely close and this causes air to leak through.[34] This leads to a breathy voice, a quality Marilyn Monroe exploited as she sang 'Happy Birthday Mr President' to JFK. This breathiness was deliberately developed by Monroe as a way to reduce a stutter, but for most other women it is not a conscious choice.

Yi Xu from University College London and his collaborators explored what qualities make a female voice more alluring to males.[35] He played recordings of female talkers to a panel of men and got them to judge the attractiveness. The phrase chosen for one experiment was 'Good luck with your exams', which does seem a little odd. Even stranger was the sentence in the second test using an artificial voice: 'I owe you a yoyo.' The results showed that on average, males preferred the female voice to have a relatively high pitch, with a wide spacing between the formants, and a breathy voice. These are all characteristics that indicate a smaller body size and youthfulness. Is the voice therefore an honest signal of evolutionary fitness? It is impossible to rule out cultural influence, however. For example, as we will see later in this chapter, there has been a gradual lowering of female voice pitch in modern times.

In men the Adam's apple is created by cartilage protrusion at the front of the throat. It is an external sign of the changed anatomy of the larynx. Some of these changes alter the motion of the vocal folds, creating an even lower, more powerful and more harmonically rich sound. The change in the vocal apparatus during puberty explains why teenage males can struggle with their voice. Popularly referred to as 'the voice break', nothing is permanently broken, however. In reality, the teenage brain is just relearning how to move the muscles that control the reconfigured larynx. Sometimes the brain gets it wrong, resulting in a voice that jumps around in pitch.

It will probably come as no surprise to read that Barry White's voice is more attractive to more women than James Blunt's.[36] Several studies have shown that on average females prefer males with a lower-pitched voice with closer spacing between the formants, something that signals a bigger body size. This is despite the fact that men with lower-pitched voices score as more untrustworthy compared to those with a higher voice![37] On average, lower voices make men more attractive to women as do more masculine faces and bodies, larger jaws, more pronounced brows, broader shoulders and being taller. This is especially true when females are looking for a short-term partner rather than someone to settle down with. In fact, the preference reflects the menstrual cycle with women preferring more masculine traits when the chance of conception is highest.[38] The voice acting as a surrogate for body size would explain why males and females have different voice pitches: it is about signalling fitness. It is important to note, however, that this is the average female response in these studies: it does not preclude some women being more attracted to ectomorphs like David Bowie or men with high-pitched voices like David Beckham.

Within genders, the voice is actually a poor predictor of body size because the vocal anatomy is flexible. An extreme illustration of this would be puberphonia, a rare medical condition where men continue to use a high-pitched voice after puberty. They hold the larynx high and keep tension in the muscles to change the vocal fold vibrations: they can end up sounding like a female impersonation in a Monty Python sketch. Underlying puberphonia is often some psychological problem that prevents the sufferer from accepting the new adult voice. Fortunately, the condition is readily treated by vocal exercises that allow the man to find his low voice. In fact, often the true voice will appear during a first session of speech therapy; it can seem like a miracle cure, as with a few weeks of practice the old squeaky voice has completely vanished.

But if the voice is not a good predictor of body size, then the female preference for a lower pitch might be signalling something else. Because the lowering of the voice is driven by the male hormone testosterone, the pitch relates to the quantity of this hormone that courses through the blood during puberty. As testosterone also influences the quality and quantity of sperm produced, is the lower voice therefore an honest signal of fertility? The answer appears to be no – because excess testosterone is harmful to sperm. There appears to be an interesting trade-off between sperm quality and the need to be more attractive to females through a lower voice. A similar trade-off can be found in other animals such as field crickets, houbara bustards and cockroaches.[39]

To get a sense of what an adult male voice would sound like without testosterone, you can look at a barbaric tradition that reached its peak in the eighteenth century. The Italian operatic superstars of the baroque period were castrati, men who had been castrated at the age of eight or nine to prevent testosterone thickening the vocal folds during puberty. They then spent their teenage years undergoing intensive vocal training. Castrati could sing at the same pitch as a boy soprano, but with the lung capacity, endurance and power of an adult. One of their virtuoso tricks was to sing a phrase for a minute without taking a breath. When a top castrato such as Farinelli performed, the audience would not shout '*Bravo*' but '*Evviva il coltello*' – long live the knife.[40]

The proliferation of castrati was triggered by Pope Innocent XI's ban on women appearing on stage in the late seventeenth century.[41] This meant the highest musical parts could only be sung by boys or men singing falsetto. But at the same time the falsetto voice was felt to lack power. During normal singing the whole vocal folds vibrate fully, opening and closing to break up the airflow from the lungs to form sound. In falsetto, some of the muscles in the larynx relax, allowing the vocal folds to stretch out and lengthen considerably. This thins the vocal folds so that only the edges are moving. And with less tissue vibrating, the pitch of the voice naturally rises because lighter matter tends to oscillate at a higher frequency – this is why a guitar has thinner strings for the highest notes.

You can find videos online that show how the vocal folds vary between normal and falsetto. The most interesting ones are recorded using a flexible endoscope that goes up the nose and then looks down on the glottis from the opening behind the soft palette. From first-hand experience I can tell you that having an endoscope up your noise is unpleasant, but it is worth it to see the vocal folds in action. They are pearly white and look like a pair of curtains flapping back and forth at the edges. For normal singing it looks like both white curtains are fully moving, whereas for falsetto only the edges are rippling. While singing falsetto, if the air pressure from the lungs is increased to turn up the volume, there is a limit to how loud the voice can get before the gently rippling vocal folds get blown open. This is why castrati were used to sing the top parts.

Becoming a castrato was seen as a chance for sons of poor families to make their fortune, and so thousands of operations were conducted each year. But while the leading castrati such as Farinelli found great fame and fortune, most were not so lucky and struggled to lead a normal life. And as the operation was outlawed by the church, it was carried out by village quacks in much secrecy, without anaesthetic and with a significant risk of picking up a fatal infection. Still, the ban did not stop castrati performing in church choirs. Fanciful excuses were made for how the spermatic cord or the entire testes got severed, such as the child being gored by a wild boar.

One of the last castrati was Alessandro Moreschi (1858–1922) who sang in the choir of the Sistine Chapel. He retired in 1912 after

Pope Pius X saw sense and properly enforced a ban on the barbaric practice. Moreschi's career overlapped with the invention of the phonograph, and so there are some scratchy wax recordings of his singing dating from 1902 and 1904. These are the only recordings of a castrato, and are easy to find on the Internet if you want to listen.[42] Moreschi's voice is in the same range as a female voice, but sounds alien to our ears. Sometimes it appears to be a woman soprano, then it morphs into what sounds like a boy soprano overstraining their voice. Baroque audiences may have revered the angelic sounds, but heard through modern sensibilities, my first reaction was one of revulsion.

Paintings of famous castrati such as Farinelli only reinforce my sense of unease. Another thing that testosterone regulates is the end of growth spurts during puberty. Castrati may have sounded like a cross between a child and a woman, but they were giants –

Eighteenth-century cartoon of an opera performance featuring two castrati (far left and right).

Farinelli was twenty-five centimetres taller than the average male at that time. Pictures of him show a pear-shaped body with overly long limbs and an expanded chest.

Modern-day scientists have tried to recreate the sound of the castrati, but are hampered by not knowing what their exact vocal anatomy was. Did they have vocal tract resonances like a adult male? This would have made a big difference to the timbre of the voice. Certainly, the tube just above the vocal folds would have been smaller in a castrato than an adult male, because its size is determined by that of the vocal folds.

Professor Johan Sundberg from the Swedish Royal Institute of Technology in Stockholm has tried to reconstruct a castrato voice for a BBC TV programme. To do so, Johan combined the sound from a boy soprano's vocal folds with the resonances of an adult baritone's vocal tract.[43] Unfortunately, as Johan described it, 'I got something rather funny-sounding' for high-pitch notes. The boy's vocal folds naturally create a sound that contains both the fundamental and a set of harmonics. A typical fundamental in the middle of the range would be 500 Hz, so in addition multiples of that – 1,000 Hz, 1,500 Hz, etc. – would be simultaneously generated. For some notes, one of these harmonics would align with one of the resonances of the adult vocal tract and a very brilliant and metallic tone would result. But for other notes, the harmonics and resonances would be misaligned, and so the vocal fold sound lacked amplification and was dull. If a castrato's voice was really like that, it would mean a simple scale would have a peculiar timbre as the power would change from note to note. Given that the castrato voice was revered for its quality, I can understand why Johan told me, with a chuckle, that 'it was a funny experiment, and I don't think I got convinced about the reality of the result'. He was happier with the simulations of the lower-pitch notes, because for those the harmonics aligned with the vocal tract resonances.[44]

This combination of adult male resonances with a soprano pitch would explain why the castrati had a distinct timbre different from modern singers. One eighteenth-century text described their voice as 'clear and penetrating as that of choirboys, but a great deal louder with something dry and sour about it yet brilliant, light, full of impact'.[45] Even modern singers with very high-pitched voice do not achieve this sound. Take, for example, the transvestite Conchita Wurst, who won the Eurovision Song Contest in 2014 with his performance of 'Rise Like a Phoenix', sporting a fine beard and a golden dress. Wurst sings in falsetto, so the dynamics of his adult male vocal folds are different from the immature vocal folds of a castrato, and this alters the vocal timbre. Even though it is high in pitch, Wurst's voice was changed by testosterone during puberty, as he became an adult.

*

Once into the second age, adulthood, the voice remains remarkably robust for many decades. There are some occupations that take their toll, though: one in five teachers miss work because vocal strain leads to them losing their voice or going hoarse.[46] But considering that the vocal folds are only one to two centimetres long and open and close a staggering 200 million times a year, it is remarkable that we have so few speech problems during adulthood.[47] From about the age of sixty, however, the voice becomes degraded due to the accumulated effects of ageing. And because speaking involves complex neurology controlling our delicate physiology, the voice provides a window on the ageing process.

Alistair Cooke was an exemplar of vocal longevity. He broadcast his *Letter from America* for nearly six decades, recording almost 3,000 episodes with the final edition going out just a few weeks before he died at the age of ninety-five. Running from 1946 to 2004 on the BBC, these weekly fireside chats provided a unique insight into American life. They also provide a rare archive revealing how one man's voice was shaped by age.

It is instructive to compare the earliest surviving *Letter from America*, which is a crackly and hissy fragment from 1947, with his final broadcast from 2004. The former was recorded when Cooke was thirty-eight and starts with the words 'A year after the dropping of the first atomic bomb ...' The last letter is an epilogue to the Second Iraq War. It may have been recorded when Cooke was ninety-five but he is still razor sharp and insightful. Yet even this great broadcaster could not completely hide the effects of age on his voice. As people get old, weakening muscles allow the vocal folds to bow and they fail to completely close. This lends a breathy quality to the voice, most heard in men, as air leaks out from the gap between the bowed folds. It means fewer words can be said with each breath, and thus older people usually have to talk in shorter bursts. If you listen very carefully to Cooke's last recording, you can detect the frequent snatched breaths between the dulcet tones. The need to breathe more also comes from a significant loss of lung capacity in old age as the cartilages between the ribs ossify and the ribcage becomes more rigid.[48] Older people also talk at a slower rate, and this too can be heard in Cooke's recordings: while the 1947 broadcast has about three syllables per second, this drops to 2.6 in 2004 – a small but noticeable change, accentuating Cooke's trademark slow, deliberate delivery.

As ageing changes the voice anatomy, it also alters characteristics such as the spoken pitch. Typically, the pitch of a male voice drops by a couple of semitones between the ages of twenty and fifty, but then it begins to rise. By the age of ninety, it is on average a couple of semitones above the pitch at age twenty. (Two semitones is the interval at the start of 'Happy Birthday'.) This rise is caused by a thinning and a change to the fibres of the vocal folds that alter their elasticity. A few years ago Ulrich Reubold and his colleagues at the University of Munich conducted a detailed longitudinal study on famous people who feature in the BBC archive.[49] When it comes to pitch, for Cooke the graph is like a hockey stick. His voice slowly drops in pitch until his late eighties, and then rises rapidly with age. His last broadcasts were delivered in the same pitch as his first, when he was in his late thirties. The researchers also noticed how Cooke's accent gradually changed over his career. For the first few decades he was nearest to General American, but later he moved closer to English received pronunciation.

Examining the Queen's annual Christmas broadcasts, Reubold and his colleagues found that her voice pitch has been dropping at about a semitone a decade over the fifty years they analysed. This is the overall trend for women, for whom the pitch decreases by a couple of semitones between the ages of twenty and eighty. But general population results have to be read with care because we don't have many recordings of the same individual through much of their adult life. Therefore, researchers have to rely on taking a snapshot of different-aged women, but here problems arise because generations may alter their voices for cultural reasons. A good example is the general lowering in women's voice pitch over the second half of the twentieth century. One study by Cecilia Pemberton and colleagues at Flinders University compared recordings of young women they took in the 1990s with archival recordings of young women from 1945.[50] They carefully matched the groups for confounding effects such as the number of smokers in each cohort. Both groups said the same phrases, such as 'The Scotch dialect is rich in reproach against the winter wind. They are all words that carry a shiver with them.' The women in the 1990s had an average pitch a couple of semitones lower than the 1945 cohort. With no obvious physical or medical cause, the most plausible explanation is a cultural shift. As we shall see in Chapter 4, females who lower their voices are perceived as being more authoritative, so this change in voice is likely to reflect the altered role of women in society.

Speaking involves exact neurological control over a large number of fast-moving muscles. As we age, there is a decrease in the number and size of fibres in the nerves controlling the larynx. This and other neurological changes affect how precisely the voice can be manipulated. Listen to Cooke's final letter in isolation and you would probably not notice these effects because everything is clearly spoken. But listen analytically and compare it to the broadcast of the first letter and you would find that the diction is not quite as crisp and the words not so precisely enunciated.

A similar change can also be detected in one of the twentieth century's most celebrated singers, Frank Sinatra. I noticed this when I analysed recordings of 'My Way'. The song was the obvious choice as it has many autobiographical elements. As Joe Queenan, writing in the *Guardian*, put it: 'What made "My Way" so affecting was that Frank Sinatra actually possessed the moral authority to sing it. A hoodlum, a boxer, a heart-throb, a has-been, a comeback kid, a titan, a has-been once again, and finally a living legend back on top for good, Sinatra had actually lived the kind of life described in the song, having taken the blows and done it his way.'[51]

The last recording of the song I found was from 1994 when Sinatra was nearly eighty. His singing indeed shows that 'the end is near', with the once great voice rendered rough by use and ageing. The singing is very clipped, presumably because of the need to take frequent breaths. What does remain, however, is Sinatra's magical trademark phrasing and timing, the way he would delay certain words for effect. Playing with our expectations of how syllables should be phrased was how Sinatra tweaked emotions.[52] But while an earlier recording, when he was in his mid-fifties, has vastly superior singing quality, for me the 1990s version carries more authentic soul, because the voice is really showing someone who had 'lived a life that's full'.

Is there anything that can be done to slow the ageing of the voice? In recent years, scientists have turned their attention to this because of our ageing population, with the number of people over sixty-five in the UK set to reach about 19 million by 2050.[53] According to one study, one in eight older people reported their quality of life being moderately or profoundly affected by voice problems. The effects include anxiety and frustration because of the need to repeat what they are saying, through to avoidance of social occasions altogether.[54]

Inevitably, people have turned to surgery to help the deteriorating voice, using a controversial procedure that has been nicknamed the 'voice lift'. The most common technique is to take fat from the stomach and to inject it into the vocal folds to make them thicker so they close better and the voice becomes less breathy.[55] However, at best the improvement only lasts for a few months, so unless you suffer from a severe vocal problem it is perhaps better to go to a speech therapist.

Researchers are beginning to gather evidence on non-invasive ways to maintain a healthy voice into old age. Like anything else involving muscles, exercise is important. To keep talking will maintain the tone of the muscles of the vocal folds and larynx. It is also assumed that this will slow the degeneration of the nerves that control the voice. It is the usual question of 'use it or lose it'. With the voice you might want to add 'don't abuse it': avoid shouting, don't smoke, and drink frequently to lubricate the vocal folds.[56] There are vocal function exercises that can be followed and have been shown to improve older voices in a small study of about twenty choral singers.[57] One such exercise is gliding from a very high tone to a very low tone on the word 'knoll' and trying to achieve this without voice breaks.[58] I suppose it's a bit like going to the gym and lifting weights: some people will enjoy the routine, but many will struggle to keep up the training as it is repetitive and boring. Such exercises are probably of more use for targeted interventions following illness or for people who use their voice professionally.

During later life older people often become isolated and therefore use their voice less frequently. Joining social groups helps with loneliness in old age, and assists the voice by getting them to chat to others. Joining a choir might be the ultimate social activity for vocal health. Singing has been shown to protect the voice from the natural decline in stability: it can help reduce warbling and enables people to speak louder, for example.[59] Good breathing, as well as larynx and vocal-tract control, are naturally trained during singing and the benefits transfer across to the spoken voice.[60] As a final incentive to join a choir, I can point to one study which shows that older singers were perceived to have younger voices than their non-singing counterparts. Joining a choir is the sonic equivalent of applying anti-wrinkle cream.

Yet even as the voice ages, the person and personality still come through. What are the features that give everyone's speech a unique character?

3 My Voice is Me

When we talk we don't just pass on words. How we speak reveals something; about who we are, where we come from and how we feel. How things are said is a vital part of communication and our vocal identity reveals intimate details about ourselves.

If your voice suddenly changes, it has a profound effect on how people perceive you, as one unfortunate Norwegian found. Germany occupied Norway for much of the Second World War and during one Allied air raid on Oslo, an unlucky resident was struck by shrapnel, blown off a road and fell down a steep incline. Astrid was Norwegian born and bred but the severe head trauma she suffered transformed her voice so she spoke like a foreigner. Her neurologist noted that 'She complained bitterly of constantly being taken for a German in the shops, where consequently the assistants would sell her nothing.'[1] Astrid had never been outside Norway, so the appearance of the accent was baffling. Unfortunately, the shopkeepers took it to be evidence that she was a collaborator. The dispassionate scientific paper on Astrid's condition reported that 'she was somewhat tearful'. This is surely an understatement: such a sudden change to her vocal identity would have been shocking and difficult to cope with.

Astrid had foreign accent syndrome (FAS), a medical condition that is fortunately incredibly rare. It arises from brain damage due to a head injury or a medical condition such as a stroke. Although people will often ascribe a particular accent to a sufferer – German in poor Astrid's case – it is usually assigned by the listener rather than being unambiguously evident in the speaker's voice. If heard alongside a native German speaker, Astrid's accent would probably not have been described as German at all; her neurologist thought it could also be mistaken for French. In sufferers of the syndrome, the brain damage affects areas responsible for controlling the vocal anatomy and this leads to faltering speech and awkward intonations. Their speech might still be intelligible, but odd pronunciations and speech rhythms give an impression of someone not talking in their native tongue. The listener then draws on stereotypes to ascribe a particular foreign accent. As we shall see later, stereotypes play a central role in vocal identity.

Nick Miller from the University of Newcastle and his collaborators interviewed thirteen people with FAS to better understand its effects on their lives.[2] Because the condition comes from a brain injury, these people are often suffering multiple problems. But for the majority their new accent was the most troubling issue. They felt that their sense of self had been altered; as one interviewee remarked, 'My old self died the day I lost the speech.' At the same time the person becomes an outsider in their own community. One sufferer had travelled all the way to a village in Poland in the hope of finding others who spoke with her newly acquired East European lilt.

Surprisingly, for some people with FAS, there are positive effects. A change in vocal persona can be an opportunity to reinvent themselves, with 'bad' personality traits ascribed to the 'old self' personified by the lost voice.[3] Similarly, though less positively, others often assume that the individual's personality has indeed been altered, even if the only real change is in the voice. One woman went from having a strong regional accent to using refined Queen's English. 'When I was in [hospital] they used to call me Miss Poshy,' she recalled. Close relationships can become strained, with one interviewee saying that her new accent was the cause of her separation from her husband: 'I was not the woman he married. I was literally a foreigner.'[4] As in Astrid's case, other people's reactions draw on preconceptions about the newly acquired 'ethnicity', and this can lead to discrimination and racist abuse. Some sufferers feel embarrassment. One interviewee who acquired an Italian accent avoided pizzerias because if the waiters 'speak to me in Italian, thinking that I am, I can't answer them back. They'll think I'm taking the piss.'

As FAS shows, listeners easily get misled by the sound of someone's voice. The brain is constantly taking a heuristic approach to learning: it tries to find patterns in what it observes in order to streamline the vast amount of information that is being collected by the senses. This simplification enables the brain to interpret information more quickly and to take action. It's easy to see why this would have been useful for survival in our evolutionary past. When you first encounter a person you tend to resort to stereotypes to judge whether they are friend or foe. What is surprising about vocal stereotypes, however, is how often the listeners get it wrong.

Take identifying sexual orientation, for example.[5] There is a style of speaking that people assume signals a gay man – we tend to think of stereotypical camp performers like Alan Carr, Kenneth Williams or Julian Clary with their high-pitched voices and swooping intonations. But confusion soon arises because some heterosexuals use this way of talking too and many homosexuals do not. Scientists have studied people's aural 'gaydar', where listeners are asked to judge sexual orientation from voices.* About 60 per cent of the time, participants in the experiments correctly identify a man's sexual orientation from the voice alone – a success rate similar to that achieved for video tests examining body movement and picture experiments focusing on appearance.[6] While listeners are doing better than chance, which would yield a success rate of 50 per cent, this still means that four out of ten times they get it wrong.

In fact, people's judgements of sexual orientation are corrupted by misconceptions. While voice pitch is assumed to be higher in gay males and lower in lesbians than it is in heterosexual men and women, the evidence reveals that this is far from the truth: the pitch heuristic that listeners use is simply wrong. Yet it is so widespread that actors play up to it when portraying gay characters on television.[7]

There are other vocal cues that together may be more useful for identifying sexual orientation among men. Scientific studies have tended to focus on the pronunciation of the consonant 's'. Stereotypically, this sibilant is often held longer, and has more hiss to it, when it is voiced by gay males.[8] The effect can be created by the tongue being either on or between the front teeth. Try moving your tongue forward and back when saying 'stereotype' and you will notice how hissy the sound becomes. It was parodied by Mel Brooks at the end of *Blazing Saddles* during the musical homage to Busby Berkeley. On stage an all-male cast in top hat, white tie and tails struggle to dance 'The French Mistake'. The outrageously effeminate choreographer Buddy Bizarre gives a demonstration. 'OK just watch me. It's so simple, you sissy Marys! Give me the playback! And, watch,

* There is a bias in the literature, with nearly all studies just reporting on male voices. This is why little information on females is given in this section.

me, faggots.' The cast responds: 'Ye*sssssssssssssssss*!' Buddy retorts: 'Sounds like steam escaping.'

John van Borsel and Anneleen van de Putte from Ghent University found that speakers with an elongated, hissy 's' were more readily identified as being gay regardless of their actual sexual orientation.[9] Since this articulation is twice as prevalent in gay men as it is in heterosexual males and females, listening out for it does increase the chances of guessing someone's sexual orientation correctly. Still, it is a very weak clue – and another reason why the vocal identification of gay men only just exceeds chance in scientific studies.

All this highlights how voice identity transcends anatomy to include social factors. It is unlikely that differences between gay and straight voices would arise due to biology: there is no evidence or obvious reason for the vocal anatomy to reflect sexual orientation. This means that differences in speech must arise for social reasons.[10] As our voice is adaptable, we can change how we speak to fit into the groups we identify with, and to differentiate ourselves from others. David Shariatmadari, a *Guardian* journalist who originally trained as a linguist, thinks that historical prejudice against homosexuals would have played a crucial role in the development of the stereotypical gay voice. A 'distinctive dialect' became a way of identifying members of the gay community that offered a safe haven in a hostile world.

In reality, the stereotypical gay voice is wrongly regarded as a female form of articulation. In the unenlightened past, this led authorities to try and make schoolboys speak in what they perceived to be a more masculine way. The American humorist David Sedaris recounted such episodes in his memoir *Naked*. He and other suspected homosexual students 'spent years gathered together in cinder-block offices as one speech therapist after another tried to cure us of our lisps. Had there been a walking specialist, we probably would have met there, too.' But this association is wrong:

in fact, a hissy 's' is not a feature of the average female voice. Shariatmadari believes this misconception probably arose from the social prejudice that gay males must be more feminine than straight men. 'The social sense that gay men are somehow more like women overlaps with that set of linguistic features,' he told me. 'Articulations like the hissy 's' 'therefore come to be erroneously identified as being features of female voices'. This would help explain why many listeners incorrectly assume that gay men have higher-pitched voices, because the most prominent vocal differentiator between male and female talkers is pitch. With more enlightened attitudes about homosexuality now prevalent, at least in the Western world, it is likely that the gay male voice is becoming less important, and that this vocal stereotyping is on the wane.[11]

*

We've all spent time preening and choosing clothes to look good. But how many of us have put a similar effort into how we sound? I guess most of us just accept the vocal identity that develops naturally. In contrast, transsexuals have to make a conscious decision to change their voice so it better portrays their gender. But adult transsexuals face the challenge of speaking with a vocal apparatus that does not match their gender identity. In recent years, there has been a dramatic increase in the number of people consulting doctors over changing gender: in 2016 there were about 15,000 gender-identity patients in the UK.[12] For those changing from female to male, high doses of testosterone can thicken the vocal folds and lower the pitch of the voice.*

* As discussed later there is more to improving the perceived gender of a voice than pitch. So speech training might also be needed to change other aspects of the voice.

The situation is more difficult for adult male-to-female transsexuals because testosterone during puberty will have already dramatically altered the vocal anatomy. This cannot be reversed by taking oestrogen. Surgery is the only way to create something physically closer to female vocal folds, but there is mixed scientific evidence on the long-term success of such operations.[13] The usual medical advice is to try speech therapy first.

We have already met Christella Antoni when I discussed stammering in the previous chapter. Christella also helps transgender clients develop a voice that reflects the person's new gender. 'I would say of all the things for transition this is probably the hardest,' she told me. 'When it comes to voice, there is no pill I can give you ... there's mostly practice and me showing you what to do.' Relearning a speaking habit is a hard and slow process, however. It typically takes six to twelve months before talking in the new voice becomes effortless. It is like learning to drive, Christella told me. Initially you are conscious of every small task you perform but gradually it becomes more automatic.

At the age of eighteen, an adult male typically has a pitch of around 120 Hz, and a female of 220 Hz. Retraining a voice to consistently jump up nearly an octave without straining is difficult. Christella reckons that about a third of her transgender clients initially display a constricted voice caused by excess tension in the larynx from ill-directed attempts to feminise the voice. And yet an overly suppressed and soft voice articulated by a male larynx can sound distinctly odd. It is better to aim for less than an octave pitch change, because this can be done with a more relaxed larynx. Scientific studies back this up: listening tests show that for male-to-female transsexuals it is often sufficient to arrive in a gender-neutral zone lying somewhere between the typical male and female frequencies.[14] Feminisation is 'all about making somebody sound authentic', Christella explained, 'which ironically is changing the voice quite a lot rather than changing it subtly'.

One of the complications here is 'response bias': even if the person's speech is mostly feminine, a single contradictory cue can make the listener notice the underlying vocal anatomy. A cough or a laugh can be all it takes for the female persona to be ruined, so the transsexual must learn to alter non-speech sounds as well. A cough immediately reveals the size of the vocal tract, which in a man will be 10–20 per cent bigger than in a woman.[15] Changing the pitch created by the vocal folds helps feminise the voice but if the resonances of the vocal tract are not also increased in frequency, then the voice may still be discerned as being male. James Hillenbrand and Michael Clark from Western Michigan University got listeners to audition samples of male and female voices that had been manipulated electronically to change the pitch and formant frequencies.[16] When only the pitch of a male voice was raised, a third of listeners still perceived the voice as being from a man. But when both the pitch and formant frequencies were raised, over 80 per cent of male voices sounded female to the test group.

Changing the vocal tract resonances is arguably the most difficult part of altering the voice. One scientific paper describes altering the oral cavity by moving the tongue forward or concentrating on spreading the lips; you can try this on the vowel 'a' to see how that alters the voice. An analytical approach based on examining frequencies is not how Christella works, however. She favours developing listening abilities in her clients to let their ears guide them to the right voice. Using recordings of their speech so they can hear what they sound like to others is vital. Because a talker's voice partly reaches the ear through internal vibrations of their skull, it tends to naturally sound boomy; in reality, therefore, a transgender person's voice might sound more feminine than they thought.

Of course there is more to the voice than a few frequencies: breathiness, inflections, intonation, articulation, loudness, even head movements and hand gestures, are some of the other traits that might need changing to feminise communication. Stereotypically, women are better at describing things than men, they use more adjectives and qualifiers – such as 'oh really' – and their voices display a more emotional tone.[17] There are many things that need to be learned by the transgender patient.

Christella is clearly doing more than just teaching new vocal skills. The voice is so central to her clients' identity that Christella has to be both counsellor and speech therapist. But not only transgender patients have an emotional investment in how they sound.

*

We all make snap judgements about people based on their voice. When I answer a phone call from a stranger, I immediately search for clues to determine whether this is a time waster trying to sell me something. When I listen to a podcast, I automatically start making assumptions about the talker's personality. Forming such judgements from a disembodied voice is common nowadays, but it was much rarer before the invention of the phonograph, telephone and radio.[18] By the end of the 1920s, however, large radio audiences were regularly tuning in to hear people they had never seen. In 1927, this led to a pioneering experiment by Tom Hatherley Pear, professor of psychology at the University of Manchester. Pear persuaded the BBC to let nine people appear on the radio where they recited a passage of prose; the *Radio Times* published a questionnaire and over 4,000 people sent in answers about how they perceived each of the speakers. In his book *Voice and Personality*, Pear explained what inspired his experiment.[19] He recounted listening to a radio play over headphones one day in a gloomy room lit only by the glow from his fire. Engrossed in the play, he conjured up in his mind what the protagonists might look like, and he began to wonder whether other listeners did the same. His interest in vocal stereotypes was also piqued by experience. 'I regret', Pear wrote, 'that for several years, misled by a voice which sounded like a carriage rolling up a loose gravel drive, I avoided making close acquaintance with one of the friendliest men in my vicinity.'

In the experiment the readers were nine diverse people who could be characters from an Agatha Christie whodunnit – they included Detective Sergeant F. R. Williams, Miss Madeleine Rée and the Reverend Victor Dams. They were asked to recite an abridged version of Mr Winkle's comical attempts at ice skating from *The Pickwick Papers* – a piece of Dickens that, as Pear described it, was at the 'midpoint of literary taste ... a passage to which the "low-brow" would not, and the "high-brow" dare not, object'. Listeners were asked to complete the short questionnaire published in the *Radio Times*, but many also sent in more detailed correspondence.

It was surprising how much people read into each voice. The most common description of the detective sergeant was of a steady and reliable man: he was imagined as being robust, of heavy build, stout, burly. But there were contrasting views:

> 'He, I am sure, is given to ranting and stirring up strife, a most bullying, unpleasant person.'
>
> 'A very human being, homely, trustworthy, conscientious ... He would have a good influence with boys.'
>
> 'A man who has little time or taste for reading, and who is employed in manual labour of some kind. He is probably well built, strong, healthy, not refined.'

Speakers on the first day (*from left to right*) : 1. Detective-Sergeant WILLIAMS ; 2. Miss MADELEINE RÉE ; 3. The Rev. VICTOR DAMS.

The mystery voices of the second day: 4. MISS A. L. ROBINSON ; 5. Captain HUMPHREY ; 6. Miss MARJORIE PEAR.

The third and last day: 7. Judge McCLEARY ; 8. Mr. H. COBDEN TURNER ; 9. Mr. GEORGE GROSSMITH.

Acknowledgments are hereby made to the following photographers:
Speaker 3, Birtles, Warrington. Speakers 4 and 7, Lafayette, Manchester. Speaker 9, Central News.

The nine talkers from Pear's radio experiment.

As we have seen, stereotyping makes it easier to interpret and respond to social situations. But while it eases cognitive overload and enables us to make rapid judgements, it also reflects racist and sexist prejudices, as a famous psychological experiment demonstrated. Developed by Mahzarin Banaji and colleagues, the Implicit Association Test measures people's reaction times to examine their unconscious bias.[20] Accurate measurement became possible in the 1990s when computers became commonplace in laboratories, and later such tests went online allowing hundreds of thousands of people to participate. In the test the picture of a person and a word appear together on the screen. The participant in the test has the task of rapidly deciding whether the word is good, such as 'love', 'laughter' or 'peace', or bad, such as 'war', 'cancer' and 'failure'. It turns out that most people, even many African Americans, tend to respond a few hundred milliseconds quicker when a positive word is shown alongside a white face than when the same word is displayed with the picture of a black person.[21] Why is there a difference in reaction times? It seems that when the combination of word and photo mirror our subconscious bias, the brain can process its response quickly because everything is as we expect. A mismatch – a combination that does not confirm our subconscious bias, such as a black face and the word 'peace' – makes participants think for a fraction longer before responding.

Stereotyping does not just serve to simplify cognitive processes. Denigrating someone outside your own 'tribe' can make you feel better about yourself and create a collective identity – like rival fans in a football stadium. But while stereotyping can lead to bigotry, without it our experience of reading a novel, going to the theatre or watching TV would be greatly diminished. Writers and actors both play on stereotypes. It allows them to create stories rich in detail – although many aspects are being filled in by the minds of the readers or the audience. When we read a book, we tend to conjure up portraits of the characters that go well beyond what is actually stated on the page. This is why we are so disappointed when we see a much-loved character from a book portrayed on screen in a way that does not reflect our expectations. This is of course unfair to the actors because they cannot possibly live up to the diverse, individualised characterisation that different readers had imagined for themselves.

But why do humans conjure vivid characterisation from the voices, as Pear found? Listening to stories helps us understand and rehearse how we ourselves should behave in social situations. The same is true when we read literature, and scientific studies have shown that reading fiction improves a person's empathy and ability to understand the beliefs, desires and behaviour of others – what is called 'theory of mind'. In some respects this is hardly surprising: we've all probably read novels that have changed our way of thinking and whose emotional content stays with us long after we have closed the book. In William Styron's novel *Sophie's Choice* a Polish mother in a concentration camp is forced to choose which of her two children will be allowed to live while the other is gassed. As we read about the harrowing dilemma, it is impossible for us not to contemplate what we might do in the same situation. It turns out that attempting to comprehend a story and trying to understand other people in real life activates some of the same brain regions. One theory is that when we read or hear a story we are running simulations of social interactions. By doing this, we learn how to behave around others, especially in scenarios that are uncommon in everyday life. In that respect, storytelling is not dissimilar from a pilot being trained in a flight simulator to deal with extreme but thankfully rare emergencies.

This simulation of life is more effective when vivid imagery is created in the listener's mind. A study conducted by Dan Johnson and collaborators from Washington and Lee University looked at the learning of emotional intelligence from fiction.[22] They asked participants to read a tale about Eric, a boy in middle school who had a difficult home life because he had an alcoholic unemployed father. His parents would constantly fight. In contrast, his teacher, Mr Howard, acted as a surrogate parent and showed compassion towards Eric. After reading the story, participants answered questions about what empathy they had towards the characters. Those participants who were trained to generate more imagery while reading felt significantly more empathy than others. They were also more likely to volunteer to help others when asked to do so immediately after the test. A vivid imagination helps us to feel more empathy and behave in a way that promotes virtuous behaviour and therefore strengthens learning.[23]

Knowing this helps explain why Pear's study elicited such detailed response from listeners to the BBC broadcast. While they only had voices to go on, they quickly sketched more complete characters in their minds whose personality traits and appearances had little or no chance of showing up in their speech. One respondent thought the policeman had weather-beaten skin, another that the judge was well dressed, and a third that the schoolgirl had blue eyes. Pear speculated that the guesses might be based on particular individuals whom the respondents had met in real life. In fact, we start learning vocal stereotypes at a young age. One study that examined role play in four-to-seven-year-olds found that the children made their voice sound deeper and spoke louder when they were playing at being Father. Some really got into character and even shouted like an angry dad.[24] But Pear's theory needs updating: we are no longer just influenced by who we meet face to face, as the voices and characters we encounter through the media also shape our vocal stereotypes.

As there are physical attributes that do alter the voice, could some of the characterisations by Pear's respondents have actually been correct? As we have seen in the last chapter, ageing affects the sound of our voice but how accurately can listeners estimate how old a talker is without any visual clues? Not very well, in fact: for much of adulthood, the voice only changes slightly, making it difficult to spot the effects of ageing. Listeners look for a variety of cues to guess someone's age, of which a slower speaking rate is probably the most useful. But beyond that, the cues they tend to rely on are mostly poor. People assume that a lower-pitched voice signals that someone is older, despite the fact that men's pitch tends to rise after middle age.[25] Features such as hoarseness, roughness and less precise articulation provide no reliable indication of age either. Still, these misconceptions may be good news for older talkers whose voices lack these features, as their age gets underestimated. For when we hear a voice that is in good condition, we assume a younger person is talking.[26]

Another characteristic that Pear's correspondents remarked on was height and weight. Miss Rée was no doubt delighted that several correspondents judged her to be slender. Can we tell height from the voice pitch? Obviously yes if we compare adults to children or men to women. But within a particular group, say adult males, someone's voice gives us no reliable cues to stature. This might explain why one respondent described the Reverend Victor Dams as 'tall' while another believed he was 'not tall'.

How might an illusory correlation around height and voice pitch arise? If we look at other species there is a general trend for small animals to have higher-pitched voices and larger animals to have deeper ones – mice squeak while lions roar. This is only to be expected, as smaller objects tend to produce higher-pitched sounds than larger items – a violin is smaller than a double bass. This relationship is also true when we compare children with adults and men with women; but again, within a group such as adult males there is no such correlation. The larynx is suspended from the hyoid bone and so its size is only loosely determined by other nearby bony structures. This means that voice pitch, which is determined by the activity of the vocal folds, does not correlate strongly with height for adult males. Despite contradictory examples, such as tall sportsmen like the footballer David Beckham or the Ultimate Fighting Champion Anderson Silva who have high-pitched voices, our brain is so keen to develop simple rules of thumb that the mind tends to dismiss these exceptions. Contradictory examples are not sufficient to break the illusory correlation because there is a strong relationship between size and frequency for many other sound sources.[27]

Have you ever had that unnerving experience of meeting someone and thinking that their voice does not match their appearance? The first time I heard the journalist and novelist Julie Burchill talk, I was surprised by her high-pitched, childlike voice. At the time she was an extremely provocative columnist whose writing set out to shock; the picture above her byline also seemed to bear no relationship to the voice. Studies examining how good people are at matching faces to voices have found that they get it right about 60 per cent of the time. While this is better than guessing, it's still pretty poor. Not only do stereotypes hinder accurate judgements: there is the added difficulty that our visual and vocal identity are mostly determined by different parts and processes in our body.

*

Vocal stereotypes also influence how we remember a voice we've heard only once. Imagine taking part in a scientific experiment where you hear a voice for the first time, and then a week later you are asked to listen to a bank of voices and identify the one you heard a week ago. When listening to an unfamiliar voice we automatically try to match it with one of a range of exemplars in our mind;[28] you might compare my voice to what you expect for the average white, southern English, middle-aged, middle-class male. In the short term, we also memorise some of the subtle ways in which a particular voice deviates from the average. But over time that detail fades and all we can recall is the exemplar. From an evolutionary perspective there is no reason why we should have more efficient ways of identifying and remembering the voice of someone we are not very familiar with: we only need to know whether that person is friend or foe. But this causes problems when voice identity is vital to solving a crime.

Dwaine George was jailed for life in 2002 for shooting eighteen-year-old Daniel Dale in Manchester. Dale was probably killed because he was due to give evidence in a murder case. One of the key bits of evidence that led to George being convicted of Dale's murder was voice identification by a witness. During the incident the killer had shouted 'you're dead now', and the witness told police that he believed it was a 'coloured person's voice'.[29] The witness was clearly referring to a stereotype and did not name George as the assailant until much later; moreover, the identification was based on having heard George talk outside a shop sometime in the previous four years. The evidence was obviously extremely weak. When it comes to voice identification, short phrases are unreliably remembered.[30] In addition, the time between hearing George talking outside the shop and the shout of the assailant would have greatly decreased chances of recognition. Even more significantly, a change of voice from an angry to normal speaking tone, like the difference between the assailant shouting and the conversation overhead outside the shop, drastically reduces recognition rates.[31] Unfortunately for George, the voice-identification evidence was admitted and used to corroborate dubious forensic evidence of gunshot residue. It took until 2014 for the conviction to be quashed, by which time George had spent twelve years in prison. For his release George could thank the students at Cardiff University's Innocence

Project, who demonstrated conclusively that the voice identification and other evidence were not sufficient to convict him of murder.

If we have problems with the voices of strangers, what about identifying a familiar voice? As my sons grew into adulthood, relatives who phoned us began to get confused between myself and the post-puberty voices of my children. When my sons were young, callers could use pitch to pick out my voice as the only adult male in the household. Now our relatives had to develop better strategies to differentiate between us – and even now they are not very good at it.[32] Usually, when someone familiar phones, it only takes a few words to recognise who it is. In Tom Wolfe's novel *The Bonfire of the Vanities*, the character Sherman McCoy accidentally dials his home phone number and gets through to his wife Judy instead of his mistress Maria:

> Three rings, and a woman's voice: 'Hello?'
> But it was not Maria's voice … 'May I speak to Maria, please?'
> The woman said: 'Sherman? Is that you?'
> Christ! It's Judy! He's dialed his own apartment! He's aghast – paralyzed!
> 'Sherman?'
> He hangs up.[33]

The ability to recognise a familiar voice is something we learn very early in our development. A foetus's heart rate increases in response to its mother's voice but slows when it hears that of a stranger.[34] Four months after birth, the baby's brain activity shows that the mother's voice is processed faster than that of a female stranger, or even that of the child's father, whose voice is still not reliably identified at this stage. For a familiar voice we memorise a few signature features that are unique to each person. The neural processing is therefore more sophisticated than that for unfamiliar voices where vocal exemplars are used. When it comes to those close to us, we need to identify individuals with more precision. We don't know which parts of the speech a newborn latches onto to recognise their mother's voice. But as it learnt the voice as a foetus listening through the amniotic fluid in the womb, it can't rely on fine details. It seems likely that pitch – both the average value and how much it varies over sentences – and the timing of the speech are important.

Some of the ability for mother and infant to recognise each other's voices must predate the evolution of speech because it happens in many other species. Picture a penguin waddling back from a foraging expedition and trying to find its mate or chick in a vast breeding colony. Penguins look and smell similar to each other and so they are very reliant on identifying relatives through calls. This they have to do in hostile conditions, with the whistling Antarctic winds and the cacophony of the rest of the colony to compete against.

Emperor penguins have particularly complex call patterns that stand out from the hubbub. Because they incubate eggs on their feet, the birds in the colonies shift about, which makes it difficult to locate mates. It gets even harder once the chicks are born and are mobile. Emperor penguins use a complex set of features to allow identification in the crowd. Birds make sound using a syrinx at the junction of the bronchi and trachea. The syrinx has two tubes but most birds only use one at a time when calling. Emperor penguins are an exception and duet with themselves by tooting both tubes simultaneously. Creating two sounds which differ a little in frequency, their calls have roughness as one tone beats against the other. In fact, the adults sound like someone doing a slow cackling through a harmonica. Scientific experiments show that emperor penguins are using a whole range of time, frequency and timbre features to make themselves recognisable to their mates and offspring. Other penguins such as the Gentoo have an easier job because they nest and so do not move about as much. For this reason the gentoo calls are simpler, sounding like party horns. Playback experiments, where biologists observe how animals respond to recordings of vocalisations, have shown that only the pitch of the Gentoo call is needed to confirm identity.[35]

Such experiments can also be conducted with humans. One study used the voices of celebrities including David Frost and Leonard Nimoy. If you play the speech of Leonard Nimoy backwards his voice is still recognisable from the distinctive timbre. With David Frost, on the other hand, the reversed speech is less distinctive and sounds more like a voice from *Twin Peaks*.[36] The characteristic timing of Frost's delivery that makes his voice recognisable gets distorted when the speech is reversed. These examples reiterate how our brains use a bespoke set of features to recognise familiar voices. By using many different attributes and individualising them for each person, our ability to recognise such voices is remarkably robust – you can still identify a loved one even if they have a heavy cold. While the human ability to recognise voices must predate speech, we are presumably even more skilled now. The number of familiar people we need to remember is far larger than for any other species, even other primates who live in social groups.

The need for robustness has in the past held back voice-recognition technologies but this is no longer the case. Banks such as HSBC started to use voice recognition in 2016 to simplify accessing accounts without people having to remember passwords and other data. Similar to the processing facilities in the human brain, a large number of features are used by the computer software to form a voiceprint that is just as individual as a fingerprint. Of the hundred or so characteristics extracted from the speech, some relate to physical characteristics of the vocal anatomy, such as the size and shape of the vocal tract, and others to behavioural traits such as speaking rate, pitch and accents.[37] Some of these must not change when people get colds, for example, otherwise the system would cease to function when you get ill. It must also not be fooled by someone imitating another person's voice. *Wired* magazine tested whether Kevin Spacey and other impressionists could fool a voice-recognition system when impersonating people like actor Christopher Walken from *The Deer Hunter*.[38] But while the impressions sounded good to the human ear, they did not fool the computer. The impressionists can copy behavioural traits such as accent and speaking rate, but however hard they might try, they cannot impersonate all characteristics defined by the vocal anatomy. Still, there are vocal doppelgängers that can trick these systems: in 2017, BBC reporter Dan Simmons showed how his bank's voice identification system could be fooled by his non-identical twin brother Joe.[39]

*

There is one voice that accompanies our lives and shapes our identity, but others cannot hear it. This is the voice you are most familiar with: the one that vocalises much of your inner speech. You will probably be using it now as you read this sentence; I was certainly accompanied by an inner monologue when writing it. Novelists often talk about the need to hear the voice of a character to be able to do it justice. In a study of ninety authors appearing at the Edinburgh Book Festival in 2014, about 70 per cent could hear the voices of their characters quite vividly. But mostly they are not talking directly to the author; it is more like the writer listening in on a conversation. As the novelist David Mitchell said of the process of writing fiction, it is a kind of 'controlled personality disorder ... to make it work you have to concentrate on the voices in your head and get them talking to each other'.[40]

The voice inside your head is not just good for reading or writing but has many different cognitive uses – it has been estimated that a quarter of our conscious waking life involves some form of inner speech.* It plays a vital part in working memory, for instance. If I was to give you my telephone number to remember, you probably would silently recount the digits in your head using the 'phonological loop' in working memory. This consists of a short-term store, which can hold auditory information for a couple of seconds, and an articulatory rehearsal process that refreshes the store. It makes use of a unique combination of speaking (inner voice saying numbers) and listening (inner ear picking up the digits). Inner speech also has an important role in motivation – for example gearing yourself up before a presentation or a job interview – and problem-solving: in scientific experiments where participants are doing an exercise while inner speech is suppressed, their performance suffers. These are all forms of deliberate inner speech, but there is also the voice that talks as the mind wanders. This is an inner monologue that does not support a specific task but verbalises thoughts. In fact, some scientists believe inner speech is a bracket term for what are in fact two different phenomena: deliberate verbalisation and daydreaming.[41]

The importance of inner speech to a sense of self is highlighted by the dramatic case of Jill Bolte Taylor, an American neuroanatomist who suffered a massive stroke that disabled the main language centres of her brain. She vividly described the moment that her inner speech started to disintegrate: 'In that moment, my left hemisphere brain chatter went totally silent. Just like someone took a remote control and pushed the mute button. Total silence. And at first I was shocked to find myself inside of a silent mind.' The complete loss of inner speech lasted for five weeks and was accom-

* Being able to think silently has obvious survival benefits – imagine trying to sneak up on an enemy where every thought that flashes through your mind is said out loud!

panied by a loss of identity. 'Dr Jill Bolte Taylor died that morning and no longer existed,' she concluded.[42]

Inner speech is not just made up of one voice, however. Take a moment to think about the voice in your own head and have a play with it. What abilities does it have? If you ask a question does the intonation rise at the end of the sentence? Maybe get it to utter the famous opening lines from *Star Trek*: 'Space. The final frontier. These are the voyages of the Starship *Enterprise*.' Does your intonation slow down to match William Shatner's deliberate delivery? Or maybe pick a character with a distinctive accent: how good an impressionist is your inner speaker?

Inner speech is very flexible and inventive. Even if you can't successfully impersonate Donald Duck out loud, your inner voice can copy some features of the famous splutterer. You can also talk to yourself with different accents. One study by two psychologists from Nottingham University, Ruth Filik and Emma Barber, got participants silently reading limericks in their heads.[43] They cleverly devised verses that would only rhyme if spoken with the right regional accent. Take these two, for example:

There was a young runner from Bath,
Who stumbled and fell on the path;
She didn't get picked,
As the coach was quite strict,
So he gave the position to Kath.

There was an old lady from Bath,
Who waved to her son down the path;
He opened the gates,
And bumped into his mates,
Who were Gerry, and Simon, and Garth.

If I talk like a resident of my home town Bristol, where the 'a' is drawn out and sounds more like 'ar', then the bottom poem works fine, but the rhyming is wrong for the last word of the top poem. If I shorten my vowels and speak with the accent of my adopted northern home of Manchester, then it's the top verse that works and the bottom one that fails. Filik and Barber monitored the eye movements of their subjects, and found that if the rhyme failed because of the accent of the inner voice, then participants glanced back and checked earlier parts of the limerick to work out what had gone wrong. This shows that while they were silently recounting the limericks their inner voice used their normal accent.

My inner voice obviously has some relationship to my external voice but the internal monologue is not just overt speech without the vocal anatomy moving.[44] This is borne out by scientific studies looking at people talking out loud or using inner speech whilst their brains are being monitored. Unsurprisingly, for inner monologues and overt talking, the classic speech regions of the brain such as Broca's and Wernicke's areas are involved. Less obviously, other regions of the brain also show activity. Recent studies show that when inner speech becomes a conversation, then 'theory of mind' parts of the brain in the right hemisphere are involved. These are networks that deal with understanding another person's perspective. It is like we are having a conversation with ourselves.

There are differences in the motor regions of the brain between overt and inner speech, as might be expected, for one involves moving the vocal anatomy and the other does not. When it comes to inner speech, additional areas of the mind are needed to inhibit the motor parts of the brain and so prevent the vocal anatomy from moving.[45] When our brain creates inner speech it needs to know that this is being generated by ourselves and not by someone else talking to us. One theory for how this is done refers to why we cannot tickle ourselves. When you command your fingers to start tickling yourself, the brain not only sends a signal to the hand, it also creates an 'efference copy' of the command. This efference copy is used by the mind to predict what sensations the tickling fingers will generate. The brain therefore has a prediction of a tickling sensation, as well as the actual sensory feedback from the skin being rubbed. If these two match, the brain knows that the sensation is self-generated and therefore inhibits the tickling sensation. A similar process could help explain inner speech: the motor signals to move the vocal anatomy are inhibited but an efference copy of the command is still created. The efference copy is used by the brain to predict what your voice would have said if the anatomy could move, and it is this that you hear. You are listening to an auralisation of your brain's predictions.

This is a neat model but it is an oversimplification. While many forms of inner speech would seem to be closely related to speaking aloud, verbal daydreaming mostly does not have an overt equivalent – an exception would be young children thinking out loud as they play. Also, inner speech is a shorthand of what might be said out loud; it appears to be more like a set of notes rather than complete speech.

Professor Charles Fernyhough of Durham University has dedicated his career to better understanding inner speech, as he describes in his book *The Voices Within*. I talked to him at the Durham Literature Festival in 2016, where he and colleagues presented their joint research. I asked Charles about how inner speech can influence identity. 'It's very strongly connected to who I am,' he told me, 'and yet at the same time it talks to me, so how the hell does that work? ... What does all that mean for the sense of self?' Scientific studies have found that the more people engage in inner speech, the greater their sense of self. And we can certainly use inner speech to change how we feel about ourselves, as this is the basis of talking therapies like cognitive behavioural therapy.

Charles also studies people who involuntarily hear voices, what in medical terms are called auditory verbal hallucinations. While hearing voices is often associated with someone having mental health issues – Charles vividly described the popular image of someone clutching their head as they are tortured by inner voices – this is a gross oversimplification. It is not just people with certain mental health conditions who hear voices: an estimated 1 per cent of the general population have such hallucinations. What underlies the phenomenon? Some believe that it indicates deficits in how inner speech is monitored by the brain. Maybe the efference message is missing, corrupted or delayed and so the brain fails to identify inner speech as being internally generated. This creates phantom voices that can have catastrophic effects on an individual's sense of identity.

Charles's project collated over 150 online questionnaires from people who hear voices. Many reported that they hear more than one voice. 'I hear distinct voices', one respondent wrote. 'Each voice has their own personality. They often try to tell me what to do or try to interject their own thoughts or feelings about a certain subject ... My voices range in age and maturity. Many of them have identified themselves and given themselves names.'[46]

The literary cliché of individuals being ordered around and tortured by voices in their head may characterise the experience of some people. But even for those individuals, there are often other voices that create positive experiences, such as acting like a guardian angel. 'I've seen people laugh out loud over funny things voices are saying to them right then,' Charles told me. This and other findings reveal that what at first might be seen as a single symptom encapsulates a range of experiences. At one extreme people hear no sound. As one respondent explained: 'It's hard to describe how I could "hear" a voice that wasn't auditory; but the words the voices used and the emotions they contained (hatred and disgust) were completely clear, distinct, and unmistakable.'

Others experiences appear to be real sounds:

> Most of the time I can hear it like it was just someone standing next to me. It's a different feeling than when you think words inside of your head; when you think inside your head your voice isn't distinct like it is when you speak out loud. You think words, not tone. But there is definite distinct tone and individuality that's unfamiliar with the voices.

If hearing voices is just a matter of the brain failing to recognise its own internal verbalised thoughts as being self-generated, it begs the question as to why there are often multiple voices with rich and diverse personas. For most sufferers, these voices are not just some form of ethereal auditory illusion, because they have enough independent agency and sense of character to be sarcastic and critical. They may be generated internally, but they can still portray an alien identity. This might be someone quite specific, like a past acquaintance or celebrity – in one study someone heard the musician Prince – but many sufferers fall back on vocal stereotypes, such as the generic voice of a policeman or a 'loud-mouthed yuppie'.[47]

Scientists are grappling with how inner voices can acquire an agency of their own.* Because external voices originate with a talker, maybe those hearing inner voices automatically assume that this speech too is controlled by someone else. Brain-imaging studies have shown that listening to a speaker activates parts of the brain linked to moving the vocal anatomy. Consequently, it is difficult for the brain to think of voices abstracted from a talker and

* A simple explanation that this agency arises because of the delusional beliefs in people with schizophrenia does not explain cases where people hear voices without such medical conditions.

without agency. For better or worse, listeners ascribe characteristics to these independent voices drawing on stereotypes.

A recent study by Charles and his colleagues has investigated the use of inner speech during reading, involving over 1,500 book lovers.[48] When reading fiction, about one in seven participants in the study recalled inner voices that are 'as vivid as if there was someone in the room with them'. One participant described the moment Lyra whispers to Will in Philip Pullman's *His Dark Materials*: 'He describes the loud, busy closeness of her whisper, and I could hear it and feel it on my neck.' But vividness of experience varied hugely, with about 30 per cent of the study participants hearing no or only vague voices. About one in five respondents had the voices from the book crossing over into their everyday lives. As one participant described it: 'If I read a book written in first person, my everyday thoughts are often influenced by the style, tone and vocabulary of the written work. It's as if the character has started to narrate my world.' As we have seen earlier, vivid imagery when reading fiction seems to aid social learning.

How do these voices sound? Commonly, readers blend vocal stereotypes and familiar voices. As Charles explained, 'If I'm reading about a woman in her seventies, and I'm given some description of her, I might blend that with the voice of my mother.' As well as drawing on known voices, people also resort to stereotypes, because these shape our reaction to unfamiliar, external voices – whether that is the Norwegian shopkeepers assuming poor Astrid is German, or the multitude of mistaken assumptions that dominated the replies to Pear's radio study.

4

Vocal Charisma

'Post-truth' was the *Oxford English Dictionary*'s Word of the Year for 2016. That year it seemed that the charisma of the messenger was more important than the veracity of any claim. While campaigning for Brexit, Boris Johnson unashamedly toured the country in a battle bus with a lie written in big bold letters down the side. The experts fumed: 'The UK Statistics Authority is disappointed to note that there continue to be suggestions that the UK contributes £350 million to the EU each week. ... [this] is misleading.' Donald Trump became US president on the back of so many whoppers that two-thirds of his statements examined by the *Washington Post* were given 'Four Pinocchios'.[1] This included the claim that he was 'totally against the war in Iraq', despite a recording of an old interview contradicting this.

Part of charisma is the aural appeal of the talker. As we shall see, there is not one definitive charismatic voice and so a good speaker adapts to the audience. Politicians are masters of this because they must adjust their approach depending on whether they address a large rally, have an informal chat with a voter on the doorstep, or give a serious TV interview. Research into charismatic voices has tended to focus on politicians, but we all use similar skills. A parent trying to persuade an intransigent toddler to share a toy, a nervous interviewee trying to land their dream job, or an irate customer trying to get a company to give him a refund: we all draw on vocal charisma.

The Greek philosopher Aristotle set out three modes of persuasion: ethos, pathos and logos. Ethos is about the credibility of the speaker, and includes characteristics such as their inclusiveness, clarity and courageousness. Pathos encapsulates whether the speaker appears to feel and echo the audience's emotions. And logos is the rational argument. In post-truth politics it seems this last tenet of persuasion has become an optional extra and bravado is winning. Worries that soaring oratory can mislead a crowd is not a new concern, however. Cicero, arguably the greatest public speaker and politician of the Roman Republic, wrote:

> Eloquence is one of the supreme virtues ... and the stronger this faculty is, the more necessary it is for it to be combined with integrity and supreme wisdom, and if we bestow fluency of speech on persons devoid of those virtues, we shall not have made orators of them but shall have put weapons into the hands of madmen.[2]

Contemporary persuasion delivers ethos through personal anecdotes, catchphrases and jokes, while pathos takes the form of emotional tales that tug at the heart strings. Even a noted orator like Barack Obama sometimes put aside logos and resorted to slogans. 'Yes we can' is a catchphrase first popularised by the children's TV programme *Bob the Builder*, and the hollow phrase worked because voters could fill it with their favourite aspirations.

When animals sing, shriek or roar, they are usually trying to affect the behaviour of others. A humpback whale sings to attract a mate, a tom cat hisses to warn off a rival, and a nightingale warbles to mark out its territory. Humans are not so different. Politicians, parents, professors – everyone uses their voice to manipulate the behaviour of others. And it's not just about the words they use. The rhythm, stress and intonation of speech, what experts refer to as the 'prosody', can reveal a lot about the talker, such as their emotional state, their personality and their origins.

One important component of prosody is our accent. It reflects our upbringing and identifies us with a particular oral tribe. Politicians often get criticised for changing their accents to fit in with their audience. During the 2015 Republican primaries for the US presidential election, Wisconsin governor Scott Walker got called out for leaving behind his classic Upper Midwest accent. As he tried to appeal to a national audience, Walker left '"Wiscahnsin" back home in Wisconsin'.[3] Such accent softening is not unusual and in the US politicians might be coached towards General American, the toned-down accent used by many newsreaders. Still, a change in a politician's voice might not be deliberate; after all, we are all vocal chameleons. When at university I shared a house with Brian from Barnsley. I remember how his Yorkshire accent strengthened whenever friends from his home town visited, rendering his speech virtually unintelligible to my southern ear. But when a politician's accent changes the public often takes this as evidence of unreliability.

Historically, many British politicians would have aimed to blend their personal accent with elements of received pronunciation (RP). RP is seen as the quintessentially British accent, although only about 2 per cent of the population use it and you rarely hear it in Scotland or Northern Ireland.[4] Typical features of RP include a slowed speech with every consonant fully articu-

---

* Except 'r's before consonants. RP is often referred to as though it is one unchanging accent, but actually there are variants and it has altered over time.

lated,* plummy 'o's pronounced with rounded lips, and the use of the broad vowel 'a' so that a word like 'path' sounds like 'parth'. Linguistically, RP is relatively young. When Dr Johnson produced his famous dictionary in the middle of the eighteenth century, he did not include advice on pronunciation because it varied amongst the educated classes. Such accord would only emerge when RP developed in the nineteenth century, arising among fashionable Londoners to signal membership of their 'higher' social group. Distinct from strong provincial accents and the cockney of working-class Londoners, it is unusual for a British accent, for while it signals social and educational background, it only gives the broadest of geographical clues as to where a speaker comes from.

RP became the accent used by the metropolitan elite, in public schools and universities, in court and on stage. Great actors such as Laurence Olivier and John Gielgud used RP when performing Shakespeare. If any regional accents were heard they would be confined to the lower orders, in comedy roles such as Bottom in *A Midsummer Night's Dream*.[5] The BBC adopted RP when it began broadcasting and it was the corporation's standard accent for much of the twentieth century.[6] The first general manager of the BBC, Lord Reith, favoured RP because it could be universally understood both at home and overseas. Only recently has a diversity of accents become common on BBC radio and TV.

Spreading around the elites of the British Empire, RP became 'the Queen's English'. This is the accent that tends to be taught to people learning English as an additional or foreign language today. Surprisingly, it also found favour in Hollywood. The cliché of the British stiff upper lip makes RP the ideal accent for a ruthless, emotionless killer – think of George Sanders playing Shere Khan in the original *The Jungle Book*. It is also the ideal voice for a baddie in a film where the little guy is fighting the establishment. Nothing says establishment so clearly as a cut-glass English accent. A good example of this is Alfred Hitchcock's *North by Northwest* where Cary Grant plays the hero, an advertising executive accidentally caught up in an international spy ring. While Grant speaks with an American accent, James Mason plays the villain Vandamm using RP.[7]

My mum Jenny speaks with a penetrating RP. If you were to meet her you might assume she was very posh. You would never guess she was born in Liverpool and only moved to the south of England after the Second World War: she has no hint of scouse. In fact, her voice illustrates the dominance of the class system in Britain. Jenny's parents were from southern England and her mother was determined that she would not pick up the local working-class accent. As my mum explained to me, 'We were all aware of social position in those days and she did not want me to talk common.' She was given elocution lessons while in Liverpool, with exercises to make sure she enunciated consonants at the beginning and end of words clearly. She had to recite phrases such as 'Five plump peas in a pea pod pressed. One grew, two grew, and so did all the rest.'[8] This was so drilled into her as a child she can still recite it from memory with perfect articulation today.

In the preface to the play *Pygmalion*, George Bernard Shaw wrote that it 'is impossible for an Englishman to open his mouth without making some other Englishman hate or despise him'. Shaw uses Eliza Doolittle's accent as a central narrative device in the play's love story. Eliza's cockney accent is transformed by elocution lessons, allowing her to gain middle-class respectability.[9] But my mum has found that having an RP accent is not always a good thing as periodically she will meet people who assume she is posh and stuck-up. One tactic she employs to counter this impression is to use what she describes as 'fairly lively language' – in other words swearing.

She is not the only one to suffer from people's wrong assumptions. Take the following exchange from Jen Campbell's book, *Weird Things Customers Say in Bookshops*:

> *Bookseller*: Would you like a bag? We've got plastic or paper ones.
> *Customer*: Well I would have asked for a bag, but you said 'plastic bag' not 'pla[r]stic bag', so now you've said that, I don't want one.
> *Bookseller*: I'm not sure people say 'pla[r]stic bag'. Also, I'm from Newcastle so I say 'bath' not 'ba[r]th'.
> *Customer*: Clearly you're uneducated.

As education tends to reduce the strength of a person's accent, this explains but doesn't excuse the customer's sentiment.

We find it easier to understand people who talk like us, and research has shown that we often prefer voices similar to our own. It's not difficult to imagine that accents once played an important role in survival. Picture a Neolithic scene where someone approaches your camp at night. Voice might be the only clue as to whether the person approaching is friend or foe. Accents and dialects afford an additional evolutionary advantage: they drive cooperation and altruism and reinforce group cohesion. This also helps explain why we are relatively blind to the variations of the voices from outside our own country. For survival it is useful to be able to identify someone as being outside our own group; more nuanced understanding of where they come from is of secondary importance.

Which sounds form accents and dialects must largely arise accidentally, for this is what happens in other animals. Many animals have accents: for example, the short thumping sounds American cod make from their swim bladder are deeper than that of their European cousins, while a unique feature of European cod is their prolonged growls.[10] Such regional variations arise when groups of the same species move apart so they are no longer conversing with each other. Then their vocalisations gradually drift apart.[11] The same thing happens in humans. Language and pronunciation are always changing. If tribes are separated and do not interact much – maybe there's a mountain range between them – then different dialects and accents become established.

Until fairly recently, it was thought that the words we use to represent nouns, verbs and adjectives arose mostly due to chance. Indeed, examining words we use for animal sounds seems to confirm that linguistic patterns for different languages are pretty arbitrary. In English a pig might go *oink oink*, but in Japanese it is *būbūbūbū*, and in French *groin, groin*. You would think that languages would use onomatopoeia for animal vocalisations and therefore similar-sounding words would arise around the world. One reason why this does not happen is that each language only has a limited number of phonemes and this makes it difficult to use words that exactly imitate some animal calls.

Generally, the sound of a word rarely has a direct relationship to its meaning. Iconic words are the exceptions, the obvious examples being onomatopes like 'ding' and 'dong'.[12] Surprisingly, a language without iconicity is a much more efficient and robust way to convey information. A recent study, however, drawing on large amounts of data, has found more iconicity than expected. Damián Blasi from the University of Zurich and his collaborators focused on the vocabulary used to refer to what is most important to humans – such as pronouns and colours and the names of body parts – and revealed that there are some underlying rules at play. Examining word lists from 6,000 languages they found that unrelated languages often use, or avoid using, certain sounds for particular concepts.[13] Take the thing sticking out from the middle of your face. In Iceland it is called *nev*, in Japan it is *hana* and for the Spar speakers in Southern Chad it is *kon*. In English, of course, we call it a nose. All these words contain the letter 'n'.[14] And there might be a reason for this. Try to say 'nnnn' but then pinch your nose while making the sound: you will realise that the sound comes entirely out of your nasal passages because your tongue closes off the mouth.

The study also found that some sounds are usually avoided for certain concepts, so the pronoun 'you' usually does not have an 'o' or a 'u' in it. You have probably done a double take reading that last sentence because this is not true for a few common languages like English. Surprising exceptions are to be expected, however, because the study is looking for general trends across thousands of languages. Indeed, the researchers found that English is often an outlier. Despite this, it seems that basic words which we all need have more commonality across languages than we would expect, including for languages that do not share a common linguistic ancestry. This commonality probably arises because the research focused on fundamental vocabulary including words that are learnt early in life. These words are more likely to have iconicity. It is only later when vocabulary expands and the brain needs to find more efficient ways to represent language that iconicity is no longer the best approach.

Britain is unusual in having a very large number of accents considering its size – it has been estimated that on average you will come across a different accent every twenty miles.[15] I live in Manchester, which is only about thirty miles from Liverpool. Yet the two cities talk in very different ways. In this case, the fierce rivalry between them is a likely reason why the linguistic uniqueness has been maintained. Just compare the Mancunian Gallagher brothers from the band Oasis to their musical idols the Beatles from Liverpool. The similarity in the musical sound of both groups is clear, yet the band members' speech is very different. Even a simple name check of their home cities reveals the difference: Liverp*iwl* or Manchest*ahhh*.

Recent research has highlighted how the map of accents and dialects in Britain is changing. Mobility is gradually turning us into verbal mongrels and it is getting harder to spot someone's origins from their voice. Adrian Leemann is a linguistic and phonetics expert from the University of Lancaster and has been working with collaborators on a smartphone app that maps accents and dialects in various countries. He had always been captivated by the differences in how people pronounce things, Adrian told me, and wanted 'to combine new approaches like app crowd sourcing with a very traditional field like dialectology'. He hoped that, eventually, these methods would allow him to reveal how universal patterns of language change across countries.

His UK app asked people how they pronounced the word 'scone'. Do you say this to rhyme with gone or stone? There is nothing perhaps more likely to cause an argument over pronunciation down the pub than that particular word. Why saying the name of a cake should have become quite so controversial and perceived to be a good linguistic marker for social class is lost in history. Maybe it has something to do with an association with rich people taking afternoon tea? Adrian and his colleagues have produced a 'scone map' of the British Isles. It shows that the north tends to favour rhyming it with 'gone' whereas the Midlands and much of the Republic of Ireland strongly favour a rhyme with 'stone'. Elsewhere, the population is split. What Adrian's map does not reveal is how the pronunciation is split by social class, but that information can be found in a YouGov survey.[16] Yet compare ABC1 to C2DE consumers, i.e. middle to working class, and you will find opinion thoroughly divided even within these social and economic groups. How we say scone, therefore, is only a very weak marker of social class; and yet our heuristic brain seems to assume that it is an excellent feature to latch onto! Whichever way you say the word, it seems people assume that the other pronunciation is wrong and signals someone of an undesirable social class, whether that is too posh or too common.

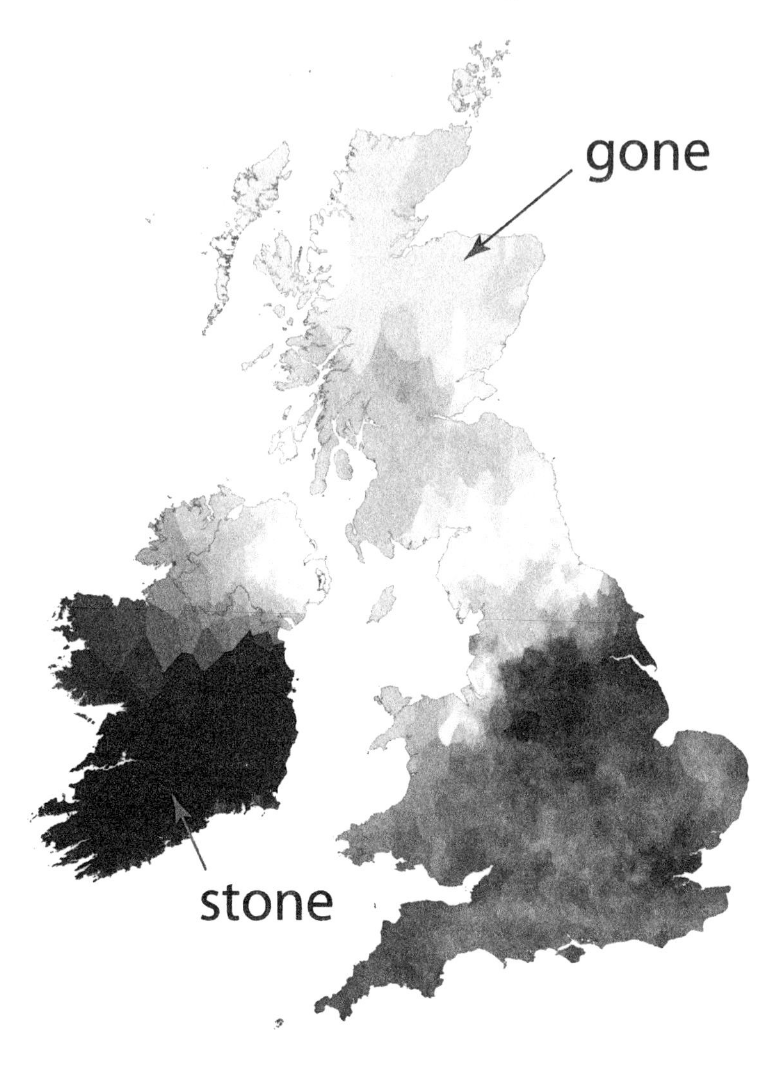

The scone map of the British Isles. Black shading indicates areas where nearly everyone says scone rhyming with 'stone', and very light shading indicates nearly everyone rhyming it with 'gone'. Mid-greys indicate areas where both pronunciations are common. (Map courtesy of Adrian Leemann, David Britain and Tam Blaxter.)

Within the app, Adrian and his colleagues ask twenty-six questions about how you pronounce certain words like scone and what colloquialisms you use.[17] They ask about the latter because differences in speech extend beyond accent to dialect – the words that are unique to a particular area. For example, nearly everyone in the British Isles uses 'splinter' to describe a small piece of wood stuck under the skin, but in the north-east the most frequent term is 'spelk'. Once the questions are complete, the app attempts to suggest where you come from. The three guesses for me were: Five Ashes, Bulford Camp and Archirondel. A fine set of English place names from the south-east, south-west and Jersey that I have never been to! Given that people find it hard to place my accent it is perhaps unsurprising that the app struggled. Most people would say I have a southern English accent. But only rarely do people spot that I was born and brought up in Bristol even though that city has a very strong and distinctive accent.

Initial findings from Adrian's research show that many local variations are sadly dying out. Dialects are more likely to change faster than accents: people stop using colloquialisms away from their home region because otherwise they cannot be understood. If when giving directions I suggested going down a ginnel, you might have a hard job understanding what I was saying unless you knew that this is the word for alleyway in Manchester and parts of Yorkshire. If on the other hand I was to use a northern flattened-vowel pronunciation for the word 'path', you would still be able to understand where to go, even if you normally spoke with a long southern vowel sound.[18]

This 'long versus flattened' vowel pronunciation is one of the few differences that still marks the vocal divide between the north and south of England. For many other pronunciations there is a trend towards speaking the English of London and the south-east.[19] Adrian and his colleagues compared their results to a 1950s survey of England.* In those days, perhaps unsurprisingly, there used to be much clearer differences between the regions. Traditionally, English had rhoticity, which means that when an 'r' appears before a consonant it is pronounced, such as the 'r' in 'arm'.[20] The 1950s maps show that the 'r' was still clearly pronounced in the south-west. Now, the twenty-first-century map shows a uniform green across all England for 'arm', indicating a loss of rhoticity.[21]

Accent matters because of the assumptions that listeners make about the talker. Speaking on BBC TV just after the 2011 London riots, the historian David Starkey provoked a storm of criticisms for these remarks:

> The whites have become black. A particular sort of violent destructive, nihilistic gangster culture has become the fashion and black and white boy and girl operate in this language together. This language which is wholly false, which is this Jamaican patois that has been intruded in England and that is why so many of us have this sense of literally a foreign country.[22]

The assumed phenomenon of black idioms being adapted by white teenagers and twentysomethings was mocked by Sacha Baron Cohen's character Ali G almost twenty years ago. Yet unenlightened parts of the media continue to complain about young Londoners talking 'Jafaican' at the expense of cockney. The way people talk is seized on as yet another example of how traditions and identity are being lost or destroyed by immigration. But this overlooks the fact that accents are naturally in flux and change all the time. Although it is easy to lampoon the new accent as 'white kids trying to sound cool', its actual origins are complex and revealing of the dynamics within inner cities.

---

* This old study only covered England and hence the lack of comments about other parts of the British Isles.

Sue Fox grew up in London and is now senior lecturer in modern English linguistics at the University of Bern. Her doctorate was inspired by noticing how the voices of the East End, the traditional home of the cockney, were changing.[23] She focused particularly on adolescent speech, because 'it is a well-known sociolinguistic tenet that "adolescents are the movers and shakers" where innovations and language change [are] taking place'.[24] Her studies involve the recording and painstaking analysis of speech from youth clubs and elsewhere.

Sue's research found that modern youngsters are clearly pronouncing the 'h' in words, for example at the start of house, something that was traditionally dropped in London accents such as Estuary. (Starkey should be pleased that this pronunciation is closer to RP!) There are many other changes in pronunciations including alterations to vowels, especially diphthongs, which are made by gliding from one vowel to another in a single syllable.[25] Sue gave me a demonstration of the diphthong in 'face'. If you say the vowels one after another, 'a-e-i-o-u', you will notice how the mouth shape radically alters to give the different frequencies of the vowels. Combine two of these together, and you get a diphthong. In cockney, there is a very audible glide when saying the word 'face', whereas in the new accent it is almost like a single pure vowel. The new accent is not better than cockney, because there is no ideal pronunciation. It is just different.

Did Sue's detailed linguistic analysis confirm Starkey's assumption that the accent is just white youth imitating African-Caribbean speaking? No, and this is confirmed by parallel changes to dialect. While the influence of Jamaican can be found in many slang terms that are commonly used – 'blood' (brother), 'yout' (youth), 'mandem' (group of friends) – there is much that comes from other cultures. In addition, there's home-grown slang, for example when Londoners talk about 'my ends' (neighbourhood). Starkey picks on aspects of the new accent that are Caribbean in origin to fit his argument, while overlooking inconvenient data that does not match his thesis.

The new inner London accent appears to arise not from white imitating black, but from a melting-pot of many accents with influences from England, Africa, the Caribbean, Asia and elsewhere. And it is used across ethnic groups. Sue first studied in Tower Hamlets where she noticed how white British adolescents would pick up and incorporate innovations from the Bangladeshis whose English is influenced by their first language. This is a two-way process, with the Bangladeshis also taking up elements of traditional cockney.

The new accent arose from extreme vocal heterogeneity, with a high proportion of people speaking English as a second language. It is significant that it originated in some of the most deprived areas of London where dense neighbourhoods and personal networks that cut across ethnic groups allowed an amalgam of inner-city voices to blend into a new accent. Something similar has happened in other cities around the world, and research projects in Oslo, Copenhagen and Stockholm are now looking at the phenomenon. It is because of globalisation and rapid changes in ethnic diversity that we are lucky enough to earwitness accent and dialect evolution that otherwise would be very slow.

Sue is passionate about trying to dispel some of the prejudice around these new accents. 'It is quite an emotive topic,' she told me, 'and there is always a perception that the language is somehow deteriorating.' She referred to the following remark by the right-wing darling of the Conservative Party, Lord Tebbit:

> If you allow standards to slip to the stage where good English is no better than bad English, where people turn up filthy … at school … all those things tend to cause people to have no standards at all, and once you lose standards then there's no imperative to stay out of crime.[26]

Sue's research into this new way of speaking aims to counter the prejudice that a particular accent indicates someone who is somehow inferior and less intelligent. In fact, it seems that people like Tebbit will just have to get used to the new voice, because there is good evidence that this is not just a short-term fad with this accent likely to be maintained into adulthood.[27] In the future, the mayor of London is likely to use this accent.

Ironically, a bigot who assumes that those who speak in an unfamiliar accent are less intelligent might be portraying their own shortcomings in being able to decode the speech. Psychology studies have shown that the ease with which a statement can be processed by the brain affects our judgement. Consider the two phrases 'woes unite foes' and 'woes unite enemies' that basically say the same thing. The first phrase is more likely to be judged as accurate because the rhyme speeds processing in the brain. The authors of a study into this, Matthew McGlone and Jessica Tofighbakhsh from Lafayette College, titled their paper 'Keats Heuristic' after the poet's famous assertion that 'beauty is truth, truth beauty'.[28] It seems that lacking anything else to go on, the aesthetic quality of the phrase makes a difference. Something that rhymes, alliterates or concisely repeats is assumed to be an aphorism and therefore more likely to be true. This is why politicians and others wanting to persuade use simple catchphrases, which by their very nature have a ring of truth about them, irrespective of their actual veracity. Advertisers play on this all the time. 'A Mars a day helps you work, rest and play', is a great catchphrase, although 'a chocolate bar a day increases your risk of obesity and Type II diabetes' might be closer to the truth.

The use of rhyme, alliteration and repeated phonemes in sound bites is so common that computer scientists can produce software that predicts what movie quotes will get onto the IMDb list of memorable sayings or what slogans will stick in the mind.[29] It has also been found that the more plosives there are in a phrase, the more persuasive it is.[30] Plosives are sounds that involve the mouth blocking the airflow from the lungs followed by a sudden release of air. 'P', 't' and 'k' are voiceless plosives and 'b', 'd' and 'g' are voiced ones – the difference between these is whether the vocal folds vibrate. Using plosives makes a slogan more rhythmic and therefore stand out more – like 'Once you pop, you can't stop' that was used in Pringles' adverts. The study also showed that using plosives can improve your retweet rate.

If our brains assume that speech which is easier to understand is more likely to be truthful, what does that say about someone talking in a second language with a strong accent? How does that accent affect the speaker's chance of being successful in politics, or more generally in job interviews or public speaking? Leaving aside even the general chauvinism that arises against minorities, it turns out that talkers with a heavy accent have a harder job of persuading listeners that they are speaking the truth, as a study by Shiri Lev-Ari and Boaz Keysar from the University of Chicago demonstrated in 2010.[31] The researchers got native and non-native speakers to read trivial statements, such as 'A giraffe can go without water longer than a camel can', and listeners had to judge their truthfulness. On a scale from 'definitely false' to 'definitely true'. Listeners scored the talker with a non-native accent lower for truthfulness.[32] (If you're curious what the answer is: the giraffe can go a little longer than a camel because it can get moisture from acacia leaves.) The results from the study imply that because it is harder to process the speech from those who have a strong accent, we have a tendency to assume it is more likely to be false. Similarly, other research has found that even native speakers with a heavy regional accent are less likely to be trusted.

Responses to regional accents are even more driven by stereotypes, however, with listeners making assumptions of social status, attractiveness and intelligence based on the voice. As we have seen, the reliance on heuristics makes such deductions dubious. In Britain, 'Brummie' is often singled out as an undesirable accent. A 2002 study found that defendants with a Birmingham accent are more likely to be rated as guilty compared to those with a more neutral accent. Circumstantial evidence of this being a learnt association comes from the most unlikely of places, the Abaco NRG nightclub, in Haifa, Israel: the club advertised in a Birmingham newspaper for staff because their clientele loved the Brummie sing-song accent. It seems that in the Middle East there are no unfair associations between the Birmingham accent and stupidity.

Thankfully, however, there has also been a general shift in perception. Many regional accents have lost their negative associations and are now often used in call centres because they are seen as warm, kind and friendly. In America, the British RP voice is now often portrayed as being particularly attractive (rather than indicating a villain). As one advert trying to attract UK visitors to Las Vegas put it, 'Visit a place where your accent is an aphrodisiac!'[33]

Our response to a voice is a mix of biological and cultural factors. Biology might explain why those who happen to have babyish voices sound less competent to us – they appear to lack the wisdom that comes with age. Cultural factors can be more difficult to determine. Some have theorised that the American attraction to UK accents comes from a desire to find mates that broaden the gene pool, but this fails to explain why some non-native accents are less attractive than the British way of talking. It seems that if there is some biological evolutionary explanation, it has long been swamped by cultural reasons. The voice helps us identify where someone comes from, and then we use stereotypes about the country to fashion our reaction. No wonder we do not like politicians being vocal shape shifters, because it prevents us drawing on stereotypes and prejudices – as unreliable as they may be.

*

Politicians, vicars and teachers all have to be good at public speaking. And in these cases, a charismatic voice is as important as good delivery and an excellent script. The ingredients of a great political speech were compiled in Aristotle's book *Rhetoric* more than 2,000 years ago, which outlined many of the dark arts of spin that people moan about in modern politics. A study of nearly 500 speeches given at British political party conferences in 1981 examined how politicians use rhetoric to choreograph the responses of party members.[34] When the politician has finished making a point, there is only a short window of opportunity of about half a second for the audience to start applauding. If the speaker signals clearly when listeners should respond, then the audience bursts into applause simultaneously. There is more to such signalling than simply pausing, however: the script needs to direct the audience.

The rhetorical device that was found to be most effective in this study was the use of contrasts, which accounted for a quarter of the applause measured.[35] Take the following example from disability campaigner Alf Morris at the Labour Party Conference: 'Governments will argue that resources are not available to help disabled people. The fact is that too much is spent on the munitions of war, and too little is spent on the munitions of peace.'

The invitation to applaud is signalled well in advance – by the phrase 'too little' – even before the last words of the sentence are spoken. Famous political speeches are stuffed full of such examples. The best make the two contrasts scan, like a piece of poetry, and so the end point for applauding is obvious. A good example is JFK's call to 'Ask not what your country can do for you, ask what you can do for your country.'

Three-part lists are another common rhetorical device, such as Tony Blair's election-winning mantra from 1997: 'Education, education, education'. It is not just used by politicians. Dickens wrote about the 'ghosts of past, present and future', the Beatles sang 'She loves you, yeah, yeah, yeah', and in *Carry on Cleo* Kenneth Williams joked 'Infamy, infamy, they've all got it in for me!' For someone delivering a persuasive speech, the rule of three provides emphasis through repetition, as well as a clear signal for when to applaud. One analysis of Obama's victory speech on election night in 2008 counted twenty-nine three-part lists in a speech lasting only ten minutes.[36]

It is often suggested that this rhetorical device is powerful because the repetition of two things can be coincidence, whereas a run of three happens less often and so we sense this might reveal an underlying truth. Our minds work hard to find patterns and make sense of the world. Brain-imaging studies, for example, have found that the prefrontal cortex examines short-term sensory patterns to work out what is likely to be heard or seen next.[37] This is a skill learnt at a very early age: if a two-month-old infant is shown alternating images to the left and right, they will start to move their eyes towards the next image in the sequence anticipating how the pattern unfolds. Similarly, part of language learning is to predict how sentences unfurl. So do three-item lists have some magical property that is ideal for our brain predictions? Studies have shown that a list of three is optimal for advertising campaigns – such as the road-safety adverts that urge 'Stop, Look and Listen' – and an additional fourth component increases scepticism.[38] We don't know, however, whether this is due to some magical completeness created by a three-item list, with our brain flagging up this length as being ideal for forming an empirical rule, or just a learnt response because we have all been exposed to so many three-part lists.

Speakers use other cues to signal to the audience.[39] The most obvious of these are hand gestures, with some speakers appearing to conduct their listeners like an orchestra. The former president of the National Union of Mineworkers, Arthur Scargill, was particularly effective, holding his hands out palm down to stop the audience applauding or moving them in a frenzy of chopping motions to signal the point where he wanted his listeners to clap.[40] But of course, how the lines are delivered, how the pitch of the voice changes, the tempo of delivery and the intonation used are also key signals for applause. And these in particular are the skills of a charismatic speaker.

Rosario Signorello, teaching at the Université Sorbonne Nouvelle in Paris, is a speech and voice scientist who specialises in researching what makes a politician's voice charismatic. He examined high-ranking politicians in Brazil, France and Italy and got listeners to rate the quality of their speeches. To remove the effects of political bias, listeners rated speakers in a country and language they did not know. Otherwise, Rosario told me, 'French people would say, "oh this is Sarkozy", and whenever I would ask them to rate his charisma they would say "oh this guy sucks" … just because he was Sarkozy.' As might be expected, the research showed the speakers adapting their voices to the audience. At mass meetings politicians use a wide pitch range to create a more engaged and lively style, whereas in interviews the pitch varies much less. Speakers tend to alter their loudness and pitch so much at rallies that it exceeds the acoustic range found in normal conversation. Signorello believes this is done to allow particular sections of the speech to connect with different parts of the audience. Those wanting their leader to show dominance and authority pick up on the passages at low pitch, whereas those looking for relatability and compassion connect with the speech at high pitches.

Of course, a barnstorming performance to whip up support at election time needs a completely different tone from a sombre conversation with fellow politicians about foreign policy. When Signorello analysed hopefuls in the 2016 US presidential campaign, he found that in the later situation candidates dropped the pitch of their voice and held it steadier. They were applying the oldest trick from the mammalian kingdom: trying to use a lower voice to portray dominance. In contrast, at a big political rally the speaker is by definition of higher social status than the audience and so they can afford to use a higher-pitched voice and vary the pitch over a bigger range. But the tone of voice also has to fit the personality of the speaker. Ed Miliband's statement 'Hell yeah, I'm tough enough', made during the 2015 British general-election campaign, failed to convince many, particularly as it came at the end of a long-winded answer.

Signorello also assessed Umberto Bossi, the Italian politician who founded the populist political party campaigning for Italy's northern regions, Lega Nord. At the beginning of his career Bossi was dynamic, authoritarian and threatening in both deed and speech, but that all changed in 2004 when he suffered a stroke. This radically changed his speech due to damage to his vocal folds. The pitch of his voice dropped dramatically by 60 Hz. He struggled to control his speech and it became breathy with a flat intonation. In experiments with test audiences comparing his speech before and after the stroke, Bossi scored similarly for competence and benevolence, but his dynamic, menacing tone had gone. His charisma had been changed by his voice, which now seems to portray a wise and calm leader.[41]

While Umberto Bossi's change in voice pitch was accidental, it is something that other politicians have deliberately done. Because the pitch relates to how the vocal folds vibrate, it is possible to retrain your voice to speak a bit higher or lower in pitch. Margaret Thatcher famously used vocal training to lower her voice to sound more authoritative, dropping her speaking frequency by 46 Hz so her voice pitch was midway between what is typical for males and females.[42]

Thatcher was not alone, however. As we saw in Chapter 2, there has been a general lowering in female voice pitch in Western countries over the second half of the twentieth century. Mary Beard, professor of classics at the University of Cambridge, has bemoaned the need for females to become 'freakish androgynes' to get their voice heard.[43] Thatcher changed her voice to make it less strident, but as Beard argues, it is unfair that a forceful woman speaking at her natural pitch should be pilloried. A man delivering the same speech at his naturally lower pitch fares much better. As Beard argues, 'When as listeners we hear a female voice, we don't hear a voice that connotes authority; or rather we haven't learned how to hear authority in it.'

This chimes with comments made by the professor of linguistics at Pitzer College, Carmen Fought, who was asked to explain why Hillary Clinton's voice had come under such scrutiny during the US presidential campaign in 2016:

> There's an idea that men and women talk differently, that men are from Mars, women are from Venus. That's really misleading. The biggest difference is in how men and women are perceived, and our ideas about how women should talk and how men should talk. Men are supposed to be assertive, loud, and competitive. Women are supposed to be soft-spoken, cooperative, and helpful.[44]

Recent neuroscience studies have supported Thatcher's approach, showing that lowering the voice pitch is effective for politicians regardless of their gender.[45] A typical experiment takes a voice recording and then uses audio-processing tools to artificially raise or lower the pitch.[46] Such tools are now widely available, with the most famous being Auto-Tune, the software that improves the tuning of poor singing. Cara Tigue and her colleagues at McMaster

University conducted a voice manipulation experiment with recordings of nine US presidents and found that two-thirds of the time the lower voice was preferred.[47] Of course the presidents were all male, but other studies have examined female voices and have found similar results.[48] Six in ten people chose the lower-pitched voices that had been dropped by about 40 Hz, the same interval as between the first two notes of the 'Smoke on the Water' riff.

A lower pitch is not only associated with the person appearing to be stronger, with greater physical prowess, more integrity and competence. As we saw in Chapter 2, a lower male voice is also seen as more attractive. So it's win-win for male politicians lowering their voice pitch because they gain dominance and attractiveness, while for females it is a higher pitch that on average makes them more sexually desirable. What a woman gains from lowering her voice is a greater portrayal of dominance, something that would have enhanced Thatcher's appeal as a leader. For women, therefore, there is a trade-off between dominance (low pitch) and attraction (high pitch).

Casey Klofstad from the University of Miami examined the 2012 US House of Representatives elections and found that candidates of both genders with lower voices on average took 4 per cent more of the vote and were 13 per cent more likely to win. This could be decisive in a close contest.[49] However, the study also found differences across the political divide, which were echoed in the 2016 presidential election: conservative voters were more likely to favour lower-pitched male candidates than liberals.

An extreme way for women to achieve a lower pitch is to use vocal fry. This is a croaking sound that turns words like 'whatever' into 'whatever-r-r-r-r-r' with a long drawn-out rattling at the end. The human voice has three registers – modal, falsetto and fry – created by different vibrations of the vocal folds. Normal speaking is done in the modal register and, as we have seen, the higher-pitched falsetto is achieved by letting only the edges of the vocal folds vibrate. In both these registers the vocal folds open and close in a simple and rhythmic motion. In fry, the vocal cartilages are squeezed tightly together, which means the vocal folds are under much less tension and become loose and floppy. This lack of tension means they open and close in a syncopated rhythm, creating a creaking sound.[50]

Early studies into women talking this way showed that the speaker was perceived to be an upwardly mobile urbanite. Culture can change the perception of vocal features, however, and this has happened to fry. This speaking style has been popularised by the Kardashians and singers such as Britney Spears – a good example of it is the opening syllables of 'oh baby baby' at the beginning of '... Baby One More Time', where each 'oh' is a drawn-out rasping sound. Now it has become hugely popular among young women, particularly as a way of marking the ends of sentences. This adoption of fry by celebrities has meant that it has also lost its yuppie connotations. Indeed, one study showed that the use of fry was harmful within simulated job interviews, with speakers using it being perceived as less competent, less educated, less trustworthy and less hireable.[51] The detrimental effect was stronger for women than men.

Mary Beard has argued that there is no neurological reason for bias against women's voices. Vocal fry supports that argument because it has been widely used by men as a sign of masculinity; a hero in an action movie like Dominic Toretto in *The Fast and the Furious* franchise will growl their way through a script. Thus women's voices are fighting prejudices learnt through culture, language and history. The preference for leaders of both genders to have voices with lower pitch is a learnt heuristic influenced by the fact that most world leaders are male. This indicates a cultural and historical bias against women leaders, as Beard suggests. While this is true, it is worth noting that this builds on misconceptions from biology that have nothing to do with prejudice. Listeners assume voice pitch gives insight into body size because it helps differentiate between small and large animals. But as we have seen, within the same gender, voice pitch is a poor predictor of human height and physical prowess. So why is someone with a lower-pitched voice seen as more dominant? As a marker for the quantity of testosterone, the lower voice could correlate with physical aggressiveness.[52]

Even if you speak at a naturally high pitch, lowering the voice for part of your speech can help signal dominance. There are many examples of animals who do this. To take one example, some frogs lower the pitch of their croaks during aggressive encounters to exaggerate their apparent size. A 2016 study looked into what humans do. It was led by the social psychologist Joey Cheng at the University of Illinois.[53] He got groups of students debating about which items of equipment were essential to surviving a disaster on the moon. It is an old psychological game, with choices ranging from a useless box of matches to life-saving tanks of oxygen. Analysis of how students interacted showed that those who lowered their voice at the start of speaking, to subtly exaggerate their formidability, were the most likely to influence group decisions.

It is curious that this desire for manliness and physical prowess still continues today. Long gone are the times where you would expect a world leader to don armour, jump on a horse and rush into battle. If the voice pitch is signalling aggressiveness mediated via testosterone, most of us would question whether it is the most important trait for a modern leader.

Looking beyond voice pitch, studies into politicians have found that faster speaking, varied pitch and changing volume all contribute to being perceived as more charismatic.[54] This is hardly surprising, because they describe someone talking in an engaged and lively manner; the droning monotone delivery used by the tennis player

Andy Murray is best avoided. The prosody of speech is one way we convey emotion, and as Aristotle wrote, a persuasive speaker needs to display pathos to appeal to the audience's emotions. Faster speaking is probably viewed as a signifier of competence because mental speed is correlated with charisma. But beware the fast-talking politician or salesman. Speaking quickly is most effective when the arguments being put forward are weak, because then listeners do not have time to analyse properly what is being said.[55]

In our social media age, empathy, informality and authenticity are arguably the pre-eminent qualities for effective communication. Neuroscience and psychology are trying to work out how authenticity is expressed in the voice, with studies showing that the way in which the pitch varies over sentences is important. In one study by Rebecca Jürgens from the University of Göttingen and her collaborators, eighty short snippets from German radio interviews were recorded that displayed strong emotions of anger, fear, sadness or joy.[56] The researchers then challenged actors and non-actors to repeat the phrases with the same emotion. The re-enactments successfully portrayed the right emotions even though the participants exaggerated changes in pitch, producing a more pronounced speech melody.[57] And it turns out that actors and non-actors were equally good at bluffing emotions. This is certainly a skill exploited by a charismatic politician, so how do we spot whether we are being duped in this post-truth era? We may find the idea of a lie detector very appealing and in Chapter 7 we will investigate whether a computer can spot untruths by listening to the human voice. But first let's look at how technology in general has revolutionised communication and changed the human voice.

5

Electrifying The Voice

Four decades after the first recording of 'Mary Had a Little Lamb', the phonograph had improved enough for Edison to host listening tests to show off how faithfully voices were reproduced.[1] At these events, a singer would stand alongside a phonograph, sometimes singing and at other times just mouthing the words while the sound actually came from the wax cylinder. The audiences were challenged to spot the difference between the live voice and the phonograph. Thousands of these events took place in front of ecstatic audiences but there must have been some bluffing involved. Surely the surface noise coming off the cylinder would have given the game away? And what about the sound being altered by the large horn needed to amplify the recording? It turns out that the singers cheated a little and impersonated the imperfect sound emanating from the phonograph. Ironically, therefore, the tests that were designed to demonstrate the machine's fidelity turned out to be an early example of recording technology changing how people sing.

To get a sense of how the voice changed in the early days of recording, I listened to a duet between Al Jolson and Bing Crosby of 'Alexander's Ragtime Band'.*[2] Bing Crosby is famous for his crooning style. You can hear a seductive lightness of tone that his microphone technique allows, especially in the second half of the recording. In contrast, Al Jolson never really adapted to the microphone and used a style that predates technology. He projects his voice as though he is trying to reach the back of a theatre, the technique of a performer from minstrel troupes and vaudeville acts. Jolson has rich drawn-out resonances with that famous twang. As his agent put it with a little poetic exaggeration, 'this man had the most resonant voice of any human being I ever knew. I stood at the back of the theatre with my hands on the wall – and I could feel the bricks vibrate.'[3]

Vocal projection was essential in the early days of recording because the phonograph lacked sensitivity. You had to sing or shout loudly into a large horn otherwise the recording was too quiet. But it was not long before the recording horn was replaced by the microphone. Singers like Crosby were then freed up from vocal techniques developed to fill large theatres and able to use a style that best suited the lyrics of the song. This was the catalyst for the huge diversity of voices that we hear in music today. But it would be wrong to assume that modern performers have simply rediscovered some long-lost natural singing voice – before they had to project their voice to a large audience – because technology has changed everything. Listen to Cher's warbling on 'Believe' or the robotic voices of Daft Punk, and it is clear that music production can greatly alter a voice. Technology does more than process what a performer sings into the microphone: it fundamentally changes their voice.[4] And this extends beyond singing: actors and actresses too have their craft radically altered when they perform on stage, screen or radio.

* Many of the musical examples are easy to find on YouTube.

The first drama especially written for radio was a play called *Danger*. Unbelievably, it was commissioned, scripted and broadcast in less than twenty-four hours in 1924. As its author Richard Hughes explained three decades later: 'Those were the days of the silent film and our "listening play" (as I dubbed it) would have to be the silent film's missing half, so to speak, telling a complete story by sound alone.'[5] Nowadays, with the wide availability of audiobooks and podcasts, it is easy to overlook how radical Hughes's concept was.

The playwright was concerned that he was introducing the audience to 'a blind man's world' and resolved to 'make it easy for them, just this once'. He did this by choosing a story where there would naturally be complete darkness. He toyed with various scenarios but dismissed a bedroom scene because he was worried about reactions from the BBC management – 'there was Major [Lord] Reith to consider'. Instead he settled on a story set in a coal mine after an accident. But Hughes reasoned that if all the parts were played by actors impersonating miners, then the audience would get confused because the characters would sound too similar. Thus the play became the story of an accident involving a party of visitors comprising two men and a girl.

Hughes wrote sound effects into the play but as the morning dawned he faced the problem of creating them. First he turned to an upmarket cinema for help because effects men were used to adding sound to otherwise silent movies. They threw peas on a drum to simulate rain, operated wind machines, or rhythmically tapped coconut shells as cowboys rode into shot. Having enlisted an effects man, Hughes's next challenge was the explosion at the heart of the story.[6] A loud bang would overload the primitive microphones and studio equipment. Fortunately the producer of the play was the resourceful Nigel Playfair. Hughes described him as 'something of a genius, and utterly unscrupulous'. Critics had gathered in a press room to hear the performance and this allowed Playfair to use subterfuge. The critics did not realise that the explosion they heard was actually created in a neighbouring room: the bang came through the wall and not via the loudspeakers in the wireless. Those listening at home got a less impressive sound, but then they weren't writing the reviews.

The final problem was the voices of the main characters. The actors were working in a studio with a dead acoustic that lacked the echoes and reverberation of a coal mine. Hughes feared that listeners would struggle to suspend disbelief if the voices did not sound right. Playfair came to the rescue again by making the cast 'put their handsome heads in buckets'. That would have altered the voice and probably made it sound like they were talking down a telephone line – a long way from the sound of voices in a mine but I guess they got away with it because of the novelty of the production. All this would be much easier nowadays of course: today a simple piece of software can add the acoustic of a mine shaft onto voices recorded in a dry studio.

To do this you need the sonic fingerprint of a mine or another space like a cave that can act as an audio double. Known as the impulse response of the space, this is what is picked up by a microphone when a short sharp sound is made in the room.[7] The impulse response and the dry recording of the voices from the studio are combined using a mathematical operation called convolution, which makes the actors' voices sound as if they are talking down the mine. This sonic teleportation is fundamental to computer games and virtual reality. Such auralisation is also becoming an everyday tool for acoustic design because it allows architects to hear what their buildings will sound like once constructed.

In radio drama, technologies like auralisation allow sound designers to alter voices of characters. But how have microphones and electronic wizardry changed acting in modern radio drama? To get some answers, I spoke to the award-winning sound designer Eloise Whitmore who had played a crucial role in one of my own research projects by creating compelling audio to demonstrate 3D sound systems – we will return to it later. Sound designers are the unsung heroines of radio drama: the hidden artists behind the sonic world listeners are immersed in. 'If the sound's really good nobody notices it, nobody says a word about it, whereas if it's bad they comment on it straight away,' explained Eloise. 'Sound design needs to fill in the picture and not overtake what's happening; it needs to be not bigger than the performance.' Subtle use of sound can help tell the story by freeing actors from laboriously recounting what is going on. To give an example, Eloise talked about the scene from *Oedipus the King* where the hero has just discovered that his wife has hanged herself. The elders tell the story and you can make out sonic snippets of what's going on in the background: 'You hear Oedipus go into the room and you hear the twisting of the rope of the body hanging. You hear him pulling the body down.' Modern digital technologies facilitate precise sound composition and enable a much more sonically rich experience than in the past. Imagine having to do this in the radio play *Danger*, with the actors standing around a microphone and a foley artist conjuring sounds from bric-a-brac.

Getting the voices right in a radio drama is more than just hiring a good actor who can say the lines in the required accent. With listeners relying entirely on what they hear, actors need to put everything into their vocal performances. As Eloise explained, 'I talk to actors quite a lot about vocalising a smile.' Emotions have to be conveyed through sounds, so a little sigh or a giggle helps to give listeners a clue about how the character is feeling. Actors also need to breathe in a slightly exaggerated way so the audience is subtly reminded of their presence. Of course, all these tiny sounds can only be picked up because of the sensitivity of modern microphones. Remarkably, breath sounds can even be used to help tell the story. Eloise worked on the police series in which Maxine Peake starred as DCI Sue Craven. This frantic drama had Craven constantly racing around the police station or rushing to a crime scene. Under the narration, Maxine's breathing tells you about how she is feeling: whether she is calm, excited or panicking. Another skill actors have to learn for radio is generating audible motion: they have to learn how to walk and talk differently to create a sense of them moving around. The quality of this sound would be unacceptable for film, but for radio it brings the voice alive.

Microphones and technology allow the listener to eavesdrop on intimate conversations, or even get into a character's mind. In theatre you have the soliloquy, but as the actor will be projecting their voice from the stage, this approach can lack finesse. In radio drama the inner voice can be much more personal. Eloise suggested I listen to a production of *Zen and the Art of Motorcycle Maintenance* she had worked on. The play follows a dad taking his son on a road trip and you hear the father struggling to come to terms with his past while wrestling with big philosophical themes. Much of the play has the dad's inner voice doing the narration, so this needs to sound different from when he talks out loud to his son. The vocal differentiation partly comes from how the actor delivers the lines but Eloise also had to help by manipulating the recording. She gave the narrator's inner voice more bass so it contrasted with the lighter external voice.[8] This matches how your voice has more bass internally in comparison to what other people hear.

Music producers also manipulate sounds to give a sense of inner voice. 'Hunter' is the first track from Björk's third album *Homogenic* (1997). The backing is a wash of synthesiser against electronic beats. Most of the time Björk's voice follows this technological aesthetic, with echo and other audio effects being added. The exception is the repeated line 'I'm the hunter', where her voice sounds natural and like a clean relay of what was picked up by a close microphone in the studio. Like the brushstrokes of a master painter, while most people might not be consciously taking in such details, the subtle production effects are a vital part of the piece. The plain acoustic of the 'I'm the hunter' line brings the singer nearer to the listener. All of a sudden Björk seems to be confessing something personal up-close.[9]

Producers are therefore playing on perceptual stereotypes even on tracks full of electronic effects. When Eloise created the voice of the Elephant-Camel, the monster in the BBC's first virtual reality film, *The Turning Forest*, she asked an actor to voice different emotions – an upward inflexion for a happy sound, a downward inflexion for a sad one, and so on. (These intonation cues are pretty universal and have equivalents in music, such as descending melodies conveying a sad mood.) Real words were avoided and a simple 'mmm' or 'huh' were sufficient. To indicate that the Elephant-Camel was moving, the actor created rhythmic breathy grunts that were followed by plodding footsteps. Then Eloise applied a vast palette of digital trickery to create the monster's sound, like lowering the pitch to create a deeper voice that fits a large beast. Listening to the final piece you would not believe it started with an actor. Nevertheless using a human voice as the initial ingredient in the sound creation embodied the monster with personality. This is another example of technology changing the actor's craft.

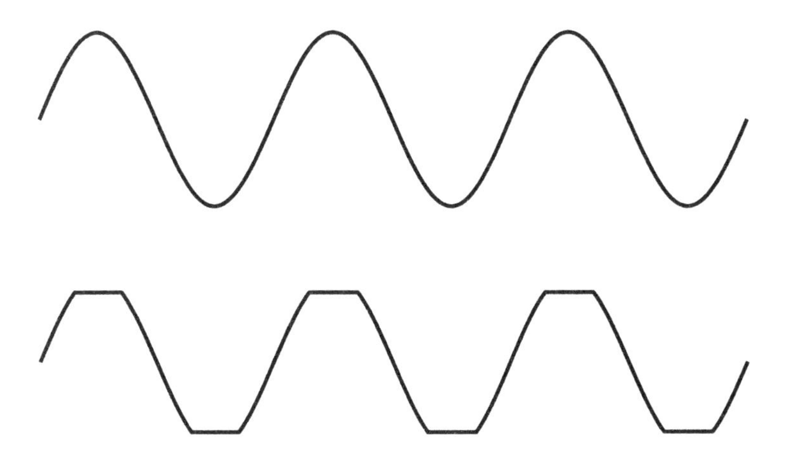

A simple sound wave and a version distorted by hard clipping.

The Elephant-Camel is a friendly monster but for a more dangerous beast a lot more distortion would have been needed. A common form of distortion involves clipping the sound wave. Imagine a typical smooth waveform consisting of alternating rounded hills and smoothly curved valleys. The strongest form of clipping cuts the top off the hills and flattens the bottom of the valleys, as shown in the diagram. The top waveform only contains one frequency, say at 100 Hz. The clipped waveform has additional overtones or harmonics at multiples of the starting frequency: 200 Hz, 300 Hz, 400 Hz, etc. While a voice naturally starts off with harmonics, distortion makes them louder and changes the timbre of the sound.

Heavily distorted vocals became a fad in the 1990s. A good example was U2's 1991 hit 'The Fly', where Bono's voice took on a gravelly tone, mimicking what Rod Stewart and Bonnie Tyler do so well naturally. Bands like Eels still do this today but not quite so severely, so that the extra harmonics make the voice sound more powerful without creating harshness. The voice sounds more powerful because the extra harmonics put more sound power at frequencies where the ear is most sensitive. The resonance of the ear canal means our ears are particularly responsive around 3,000 Hz. Add even more distortion, like U2 did for 'The Fly', and this results in harshness, or what scientists would describe as 'roughness'.

Within the inner ear the sound is split up by the basilar membrane according to frequency. The membrane stretches along the length of the cochlea; if you uncoil it, it's a bit like a piano with the frequencies lined up along its length. High frequencies excite the basilar membrane closest to the oval window and low frequencies move the membrane at the opposite end. When listening to a sung note, the basilar membrane will be simultaneously vibrating at specific points and this signals to the brain which frequencies make up the sound. For low-frequency harmonics the vibrating parts of the basilar membrane are far apart. In contrast, for higher frequencies the harmonics are so close together that the vibrations of the basilar membrane created by the harmonics interact with each other and the movement becomes complicated. When that happens, the brain hears this as roughness.

Roughness happens naturally during a scream. Imagine starting with a quiet 'aaa' sound and gradually increasing the pressure of the air flowing from the lungs. Initially as the sub-glottal pressure driving the vibration increases, the vocal folds move more and this creates a louder sound. But if you really let rip, the physical limitations of the larynx are reached and the vocal folds can no longer flap open and close in a simple regular pattern. As a result, the voice gets distorted and the squealing high-pitched 'aaa' acquires a screeching roughness.

In 2009 I ran an online experiment for the Manchester Science Festival where people were asked to rate the scariness of nineteen different screams. Editing the sounds for the experiment was very unpleasant because the person screaming sounded very distressed in the most potent examples. What made them sound so horrible was the roughness, which only happens when someone screams uninhibitedly and with maximum force. An analysis of 20,000 ratings found that female screams tended to be scariest, with the longest and highest-pitch screams being the most blood-curdling. Women's screams are naturally at a higher pitch than men's and so are closer to the frequencies where people's ears are most sensitive and therefore sound louder.[10]

Underlying the roughest sounds is a rapid fluctuation of the sound wave at around 70 Hz. Intriguingly, this is a niche that is not used by normal speech. It sits between the rate at which the articulators of the mouth move to shape words, and the frequencies within the buzz made by the vocal folds. A distress call often requires immediate action to prevent danger, and so there is an advantage in having this roughness as it makes screams stand out from speech. Luc Arnal from the University of Geneva and his team have investigated this.[11] In one study they showed that adding roughness to a sound not only increased perceived fear, it also prompted subjects to respond quicker. By placing the subjects in an fMRI scanner the scientists could also examine how the brain responded. The unpleasant sounds largely affected the amygdala and the primary auditory cortex. The amygdala is a collection of neurons that are almond-shaped. Located low down in the brain, it is known to play an important role in helping us detect and deal with threats.[12] The harsh sound that can be created by heavily distorted voices seems to be playing on cognition that evolved to allow us to quickly identify and react to distress signals. No wonder distortion is popular with heavy-metal singers – and useful for creating malevolent monster sounds.

The Turning Forest, the virtual-reality film that featured Eloise's Elephant-Camel, had an unusual incubation because the soundtrack was created before the images evolved: usually audio is subservient to the visuals. The soundtrack was actually created for a large research project that I'm part of that is developing new ways to hear audio in the home.[13] To investigate the limitations of current technology, we commissioned a series of drama vignettes. The brief for the scriptwriter was odd, with a table of technical requirements that showed little interest in producing a good story! We requested a first-person perspective, with sound sources moving, approaching from various directions and heard at different distances. We did this because we knew that current audio systems struggle to create such a diversity of sounds. Out of this strange brief, the writer Shelley Silas managed to conjure a magical fairy tale with a boy interacting with a huge Elephant-Camel trampling through a forest. It was only later that the BBC commissioned the visuals that turned *The Turning Forest* into a VR film.

Hopefully one thing to come out of the research project will be ways to improve the intelligibility of dramas on television. In recent years there have been a whole string of TV programmes that drew complaints as viewers struggled to hear and comprehend the dialogue. (You do not get this problem with radio drama, because if words become unintelligible the broadcast becomes pointless.) The costume drama *Jamaica Inn* earned the nickname 'Jamaica Inaudible' after it attracted thousands of complaints. A more recent example was *SS-GB*, a what-if drama exploring Britain after a Nazi victory in the Second World War. The lead actor was Sam Riley and his voice was a barely audible rasp at times. Maybe some scenes should have been prefaced with 'Listen very carefully, I shall mumble this only once.'[14] Sound recordists have rightly been annoyed when they were accused of not doing their job properly. In fact, the voice was captured faithfully with extremely sensitive microphones. The problem is that these microphones allow actors to perform more naturalistically without projecting their voice: thus the whispering is a stylistic choice by actors and directors who favour naturalism over diction.

The research project is looking at other common intelligibility problems, such as music being too intrusive and making dialogue hard to hear. The solution to this may come from object-based audio. When you watch TV, you typically get two streams of audio from the broadcaster that are sent to the left and right loudspeakers in your television. If the music is too loud then it is difficult to turn it down because the speech and music are already mixed together. In object-based audio the music, sound effects and dialogue are all sent separately to your home. Here a bespoke mix is produced by your television, which allows you to turn down the music if that's what you prefer. At present my colleagues are developing computer algorithms to monitor how intelligible the words are in a particular scene, which then allows the television to automatically adjust the volume of the background sounds to make sure the words never get swamped.

Technology has also created intelligibility issues in live theatre with TV and film influencing the aesthetic of playwriting, acting and production. A desire to move away from someone blasting their voice out in received pronunciation to more naturalistic speaking is one of the issues. But if a theatre director decides to use authentic accents that are harder to decode, how are people in 'the gods' going to be able to hear? The solution would seem obvious: put a microphone on the actor and some loudspeakers pointing at the audience and turn up the volume. This is how they do it on Broadway where the audience expects actors to be amplified even for speech theatre. But in the UK overt use of electronics is controversial.

One high-profile example was the outcry at the National Theatre in 1999 when it was revealed that electronic reinforcement was being used for Shakespeare. As Graham Sheffield, then the artistic director of the Barbican theatre, commented, 'Using microphones for special sound effects is one thing, but it is quite another if they become a general support mechanism for lazy actors.' Sheffield went on to add, 'It does destroy the intimacy and naturalness between actors and an audience. However well it is done it will always sound slightly artificial.'[15] Interestingly, complaints only started a few months after *Troilus and Cressida* had begun its run. The critics and thousands of punters had already seen the play with electronically reinforced voices but no one had noticed. Indeed, the show had garnered good reviews, with Michael Billington in the *Guardian* calling it a 'magnificent new production'.[16] The use of electronics only hit the headlines because someone at the National Theatre leaked the facts to the press.

I first heard about this controversy when I saw a presentation by the theatre sound designer Gareth Fry at an audio conference in 2010. Backstage creative wizards such as Gareth are not well known to the public, but his credits include the sound for Danny Boyle's Opening Ceremony of the 2012 Olympic Games in London.[17] More recently, I met up with Gareth at HOME theatre in Manchester, as he took a break from a show he was working on. When I asked him to explain what his job entails, his brief answer was, 'I'm responsible for everything the audience hears.'

Gareth explained that the sound issues at the National Theatre were an accidental by-product of changing fashions in staging. When the theatre was built most shows had large, heavy sets. A Noël Coward play like *Relative Values* would have a very naturalistic rendition of a library in a grand house beautifully painted in minute detail. Such sets cannot easily be moved and this means that all the plot developments must occur in this one room. The writer has to contrive reasons for people to visit the room and the audience needs to suspend disbelief when even the most unlikely encounters happen in this one space. The fixed set might cause challenges for writers but it has a big acoustic advantage: sound reflects from the heavy scenery into the audience and so reinforces the actor's voice.

By the late twentieth century, set fashion had changed, however. Following the aesthetic of TV shows and movies, writers wanted to be able to switch location. This meant scenery had to be simpler, lightweight and abstract. A move to a new location is often achieved by simple changes to lighting and the soundscape. This means that the useful reflection of sound from the scenery has disappeared; and without the big heavy sets some members of the audience get poor sound. It is to counter the deficient acoustics that electronic assistance is needed – not because modern actors are unable to project their voices, as some journalists have complained. It has become even more important nowadays when effects and music are added to a play so actors' voices are competing with more 'noise'. But this must be done subtly so the audience is unaware of the reinforcement. Gareth described the technique as 'halving the distance: just about making the performer seem like they're half as far away as they are'. But the use of technology can go further to aid storytelling. Gareth talked about the electronics acting as an auditory mask, 'dislocating a voice from a performer'. Simple processing like pitch shifting can be used to change a character from male to female, and reverberation added to change the actor's location from a bathroom to a church.[18]

In music, reverberation is widely used to improve the voice. It is a kind of auditory ketchup because adding a bit to a recording usually improves it.[19] When music producers do this they play on listeners' expectations and stereotypes. Patti Page's 1948 record 'Confess' was groundbreaking because it was the first hit where a pop singer was overdubbed: Page accompanies her own singing. It is a call-and-response song with reverberation added to the second voice. It was achieved by a loudspeaker playing Page's singing of the second line into a men's room with a lively acoustic. The reverberated sound was then picked up on a microphone and recorded. Adding reverberation differentiated the call-and-response lines that otherwise would have blended together because they were both sung by Page. Reverberation has religious connotations because it naturally arises from the acoustic found in churches and cathedrals – very appropriate for a song called 'Confess'.[20]

I talked to Gareth about the play *Encounter*, a worldwide hit performed on Broadway in 2016. Unusually, sound dominates both the staging and storytelling of the play. It tells the true story of Loren McIntyre, a photographer who got lost in the Amazon rainforest in the 1960s and found shelter with the Mayoruna tribespeople. It could have been presented as a traditional piece of theatre, but as Gareth pointed out, recreating a rainforest with scenery and projection was problematic. The play would be 'immediately set for failure, because it was always going to be a diminished version of reality', he explained. Much better to draw on the audience's imagination through the use of sound and let them conjure the scenes in their minds. This went further than just playing a rainforest soundscape over loudspeakers, however: each member of the audience was wearing a pair of headphones thus enabling better storytelling.

Encounter contains many overlapping narratives, including a father telling bedtime stories to his daughter. In a normal theatre production it would be hard to realise that intimate moment fully because of the physical distance between stage and audience. But by giving every member of the audience a pair of headphones connected to a special microphone on stage, it was possible for the actor Simon McBurney to whisper into the audience's ears. Towards the start of the performance, McBurney blows across the microphone and a squeal rises up from the audience as it feels as if their ears are being warmed by the actor's breath. Here the one-to-one intimacy of being read a bedtime story is recreated. The audience is aurally brought on stage and transported to the rainforest.

The special microphone on stage was a dummy head, which is a manikin with microphones in the ear canals. Gareth had also used it to record in the Amazon rainforest. It was a challenging recording trip because of the 'fricking mosquitos',[21] he told me: he 'couldn't bat them off because it would've ruined the recording'. The charcoal-grey disembodied dummy head captured the sound in binaural, a technique that is a mainstay of acoustic research.

A dummy head.

Close your eyes and listen to the sound around you. There might be a car in the street, in another direction a bird is singing, while elsewhere you might hear a distant radio. The aural clues that tell you where the sounds originate from are overlaid on the sound waves travelling up the ear canals.[22] The reason a dummy head has microphones in the ear canals is to capture the sound waves with all the spatial cues. If you then reproduce the recording on headphones straight into the listeners' ears they are being aurally transported to where the recording was made. Indeed, a good binaural reproduction will make sounds appear to be coming from outside your head. This is different from a normal recording: listen to an average music track over headphones and the band appears to be performing from inside your skull. This happens because the acoustic cues that would place the musicians outside your head are missing. The brain cannot work out where the sound is coming from, and reasons it must be coming from within. Gareth played with this when staging the story in *Encounter*. The internal monologue of the lost photojournalist is played in normal stereo and so appears to be inside the heads of the audience, while the members of the Mayoruna tribe he encounters are rendered in binaural and their lines seem to come from outside the listeners' heads.

The use of binaural technology has until recently been largely restricted to laboratories. But elsewhere it is undergoing a revival because headphone listening has become hugely popular. The BBC have broadcast an episode of *Doctor Who* with a binaural soundtrack, 360 videos on the Internet use the same technology, and it is how sound is reproduced in VR headsets. So as storytelling adapts to virtual and augmented reality, how might technology change the actor's voice?

*

Before recording technology, actors and singers in large venues faced an athletic challenge: how to reach people at the back of the theatre with more than a disappointing murmur. This partly explains why singing styles such as opera sound so peculiar to many modern ears.[23] A fascinating contrast of the old and new can be found in 'Barcelona', the 1988 duet between Queen's extrovert rock frontman Freddie Mercury and the operatic soprano Montserrat Caballé. The song gets interesting after the overblown and lengthy introduction, with its pastiche of rock and classical music and its bombastic chords, timpani rolls and tubular bells. Mercury's voice is very expressive: many would say that he was one of the greatest singers of the twentieth century. Sometimes it has a sweet melodic tone; at other times he is almost shouting as the lyrics ring out with energy. It was microphones and amplification that allowed

Mercury to express such a wide range of emotions even when singing to huge audiences. In contrast, Caballé's voice is soaked in vibrato and always very melodic, sounding almost like a musical instrument. But while, following the operatic tradition, the quality of Caballé's vocal timbre is paramount, the articulation of the lyrics is less important.[24] Indeed, at times it is difficult to work out if Caballé is singing in Spanish, Catalan or English.[25]

The vocal gymnastics that Caballé employs are certainly dramatic but they come at a price that goes beyond unclear lyrics: the singer has a restricted vocal palette to convey emotion. She can play around with timing, harmony and dynamics as any instrumentalist does, but there is limited scope for crooning into the note or other stylistic changes. Caballé cannot do what Mercury does and radically alter stress and intonation to bring out her personality. Thus, to the untrained ear, operatic sopranos all sound very similar. In contrast, the best pop singers are often those with the most distinctive voice. Take Bob Dylan, for example. When he toured in the mid-1960s his use of an electronic backing band caused uproar among fans. At one infamous concert at the Manchester Free Trade Hall there was slow hand-clapping and one of the audience shouted 'Judas'. Yet to accuse Dylan of selling out his authentic folk roots is ironic, because without a microphone and amplification, he would never have been able to sing to large audiences with his distinct gravelly delivery.[26]

I recently sat in the front row of an opera recital and felt the full force of the operatic voice. The pianist was forced to hammer the keys in an effort to make the accompaniment audible over the loud singers. Vocal power is vital for performing grand operas because singers are usually competing with an orchestra, sometimes a huge group – that for a performance of Wagner's Ring Cycle can have ninety members. The fact that the orchestra is playing in the pit helps the singer to be heard above the instruments. In a theatre such as Wagner's Bayreuth Festspielhaus nearly half the orchestra is located underneath the overhang from the stage. If the direct path from a musician to the listener is blocked, then the sound is only heard because of a reflection or because it bends around the pit rail due to diffraction. Low frequencies diffract more easily than high frequencies and this dulls the orchestra's sound. There is less aural competition for the singer in the higher bandwidth. But even with this help from the theatre pit a specialist singing technique is needed.

Opera singers target a frequency range where the ear is particularly sensitive. The ear canal between the pinna and eardrum has a resonant frequency at which the air in the tube vibrates efficiently. This resonance means that anything the singer produces around 3,000 Hz will naturally sound louder because of the anatomy of the ear. To reach that range, male and female singers have to adopt different techniques, however, because their melodies are at different frequencies.

Take a male baritone singing a low note at 100 Hz, a long way below the most effective bandwidth. This note will also have harmonics that are multiples of the note frequency – 200 Hz, 300 Hz, 400 Hz, etc. To amplify his voice, the baritone tunes the resonances of his vocal tract to one of the higher harmonics in the frequency range where the listener's ear is most sensitive. This creates what is called the singer's formant. The baritone achieves this by lowering the larynx and narrowing the vocal tract just above the glottis. Actors do something similar to project their voice.[27]

Sopranos have to adopt a different approach because they sing at a higher frequency (300–1,000 Hz). They tune their vocal tract to follow the fundamental pitch of their vocal folds. They do this by opening their mouth wide so the vocal tract gradually flares like a megaphone. But as they soar for high notes they run into problems because some vowel sounds are then impossible to create faithfully.[28] This is one of the reasons why it is hard to work out what words Caballé was singing. Pop music, on the other hand, uses amplification and so allows lyrics to be understood when performing to large audiences. Indeed, the lyrical inventiveness that is a hallmark of many great pop ballads is only possible because of microphones.

*

Modern singing has gone through many iterations driven by both culture and technological innovation to give us the diversity of voices we hear today. Arguably the first and most important was singing in a more natural way. The introduction of microphones meant that singers could perform in a conversational manner. This new singing style, called crooning, is associated with American artists such as Bing Crosby, but the first crooner is generally thought to be Al Bowlly who was born and brought up in Africa.[29] Bowlly came to Britain in the 1920s where he stayed for much of his performing life, until his tragic death in a bombing raid during the Second World War. Watching an old film clip of him performing 'Melancholy Baby' in the British Pathé studios shows him singing to a large microphone on a stand. It looks almost as if he is addressing, and singing sweet nothings to, a melancholy lover sitting in a chair right in front of him. Bowlly leans forward and almost whispers the most intimate lines into the microphone, before leaning back and singing more conventionally such upbeat lines as 'Every cloud must have a silver lining'. His light tenor voice is extremely precise, allowing for subtle changes to bring out the meaning in the lyrics.

Crooning sounds very quaint to our ears today, but when it first appeared this public display of intimacy attracted great controversy. 'I cannot turn the dial without getting these whiners and bleaters defiling the air and crying vapid words to impossible tunes,' complained Boston's Cardinal O'Connell in 1932. 'Crooning is a degenerate form of singing,' he went on. 'No true American would practice this base art.' In Britain Cecil Graves, controller of programmes at the BBC, issued instructions to keep 'this particular form of odious singing' off the radio.[30] Critics saw crooning as being effeminate and emotionally inauthentic, but they lost the argument.

One of the greatest crooners was Bing Crosby. Beginning his career as a vaudeville act, he readily adapted his stagecraft to the new possibilities offered by the microphone. Crosby's most significant contribution to the development of music did not come from his singing, however: it came from his bankrolling the development of tape recording. The introduction of magnetic tape improved the quality of recorded sound and, even more importantly, allowed easy editing using just scissors and sticky tape. Mistakes during a performance were no longer indelibly etched into wax or resin but could be removed.

Crosby hated having to repeat live radio shows so they could be broadcast to the different time zones in the United States, as it reduced the time he could spend on the golf course. In 1946, the fledgling ABC Radio Network tried making life easier for the megastar by pre-recording his show *Philco Radio Time* on disc. But the sound quality was terrible and the audience quickly realised that Crosby was not performing live and ratings suffered. The solution came from recently defeated Nazi Germany. Tape recording was invented and used in German radio broadcasts during the Second World War; in fact, the first indication that they were broadcasting recordings came when the Allies heard orchestral music being played in the middle of the night when musicians would be asleep. They also knew the Germans used something better than cylinders and discs because the broadcasts lacked the characteristic surface scratch and crackle. After the war, the reel-to-reel tape recorders called Magnetophons were discovered and shipped back to America where the new technology was dissected, copied and improved.

Crosby realised the potential of magnetic tape to make his life easier, and funded the development of the technology.[31] Once magnetic tape was used for his show, Crosby could almost whisper into the microphone as these quiet sounds were no longer swamped by surface noise from disc recording. The listeners thought he was performing live again and the ratings for his radio programme recovered.[32]

Others realised the potential of electronics to aid the expression of raw emotions in songs that reflected real-life struggles. Billie Holiday was one of the greatest jazz singers of the early twentieth century. She had a troubled childhood and made a living by scrubbing floors and running errands for the girls in a brothel before turning to music to make money. As her obituary in the *New York Times* put it: 'Miss Holiday became a singer more from desperation than desire.'[33] Listen to her singing 'Strange Fruit', a harrowing depiction of the lynched bodies of black men and women strung up in poplar trees, and you feel that she is drawing directly from the trauma of her own life. Her voice has a disjointed tone with a sorrowful delivery that would have been impossible to deliver if she had been singing loudly without microphones.

Present-day singer-songwriters too make the listener feel that they have access to the musician's inner psyche. As the music journalist Kitty Empire wrote, 'music fans thrill to the spectacle of a singer-songwriter's inner torment. We think songs make a direct connection to the artist's tenderest places. We are rapt by the crack in a voice, the glint of a tear.'[34] Professor Nicola Dibben researches emotion in music at the University of Sheffield, and has not only written about intimacy in recordings by Amy Winehouse and Adele but also worked with Björk. She told me that the evolution of recording technology, and especially the use of the microphone, has created 'a shift to a very individualistic and almost unhealthy relationship [of the audience] with individual stars'. Close-ups in film helped create the cult of the movie star; the close-up voice picked up by the microphone naturally fosters pop stars.

Nicola finds inspiration for her research in the most unlikely places. Getting her hair done one day she was taken aback by the reaction of her hairdresser to a song by Adele. Normally the background music in the salon is largely ignored but when a track from Adele came on the hairdresser said things like 'oh my God I love this track', 'she talks about things just like I experience them' and 'it's though she's speaking about my life'.

Nicola examined Adele's hit 'Someone Like You' to investigate how music producers enhance intimacy in pop music.[35] The song was the top-selling UK single in 2011 and won a Grammy for Best Pop Solo Performance. The lyrics are very emotive, telling an autobiographical story about coming to terms with the end of a relationship. Adele sings against a simple piano accompaniment, which gets increasingly intense over the course of the song. But the key to making it sound intimate is to make you feel that Adele appears to be singing *physically* close to you. Naturally, having another person extremely close by heightens our emotional response. To achieve this the music producer places the microphone close to the singer and then applies an audio effect called compression. This boosts the quietest parts of the singing so that tiny sounds like the performer's breathing become audible.[36] It involves a great deal of artistic subterfuge because the compressed singing is not exactly what you would hear if Adele sang without a microphone. While in the early days recording tried to faithfully capture a singer's voice, over the last fifty years producers have endeavoured to enhance it. Thus even what appears to be a stripped-back track such as 'Someone Like You' is actually a hyper-real production. Like many other aspects of modern sound design, however, this only works if the listener is unaware of the sonic trickery.

Adele's voice in a ballad like 'Someone Like You' needs compressing so that she can use her full vocal abilities. If you juxtapose the opening and end of the track then the huge contrast in singing between the contemplative start and Adele letting rip at the end is obvious. Yet the final part of the track sounds only a little louder than the beginning, because compression has been applied to the opening.

This resulting sense of intimacy can also be pretty creepy, Nicola Dibben explained that if you subject Jarvis Cocker's incredible vocals to cold analysis it reveals them to be 'really repulsive actually'. The Britpop albums that Cocker recorded with Pulp in the 1990s deal with thwarted romance, voyeurism and sex. Singing about an affair in 'Pencil Skirt' Cocker's vocals are full of exaggerated lip, tongue and breathing sounds. This aural close-up puts the listener in close proximity to the singer. The voice turns the listener into a voyeuristic accomplice in the affair.

While Jarvis Cocker and Adele cleverly play on the intimacy offered by the microphone, there are those who perhaps take it too far. Stars such as Mariah Carey and Whitney Houston, as well as many contestants on programmes such as *Pop Idol*, have been accused of 'over-souling' – an excessive reliance on verbal gymnastics and over-emoting. In some cases the result is comical. Christina Aguilera's rendition of the US national anthem at the Super Bowl in 2011 is worth looking up online. As John Eskow wrote on the Huffington Post site, 'singers like Aguilera – who admittedly possesses a great instrument – just don't seem to know when to stop, turning each song into an Olympic sport as they drain it of its implicit soul, as if running through the entire scale on every single word was somehow a token of sincerity'. The desire to excel on TV talent shows, and the need for a song to keep listeners engaged in a world full of distractions, has led to over-the-top performances that I find exhausting to listen to. But maybe I am like those who complained about the early crooners and it is a sign of me getting left behind by fashion.[37]

*

When we think about how the voice has been changed by technology it is important not to overlook an aspect more loosely connected with technological innovation – the influence of mimicry. The availability of vast music libraries such as Spotify to budding singers is changing the voice. There are wannabes listening over and over again to recordings to hone their voices to match one of their vocal heroines or heroes.

Helena Daffern at York University is both a professional singer and a researcher experimenting on the voice in her laboratory. When we talked about mimicry, Helena pointed out that today people are trying to impersonate recordings where voices are coated with audio effects. 'How has that changed how we sing?' she asked rhetorically. 'When you're a kid and you're singing along to Beyoncé you're not waiting for a producer to put all those effects on, you try to make them yourself.' But it's not just effects: people will be trying to mimic singing that has been made more perfect through editing. Unless they have undergone intensive training singers are unlikely to immediately hit the right frequency for each note. When a recording was etched into wax such faults were there forever but with digital technologies these 'flaws' can easily be removed. A singer will often record a vocal half a dozen times and then the music producer will cut and paste together the best bits from all the takes.

At the same time, bum notes can now be corrected by software. Auto-Tune is the aural equivalent of digital photo editing that enables you to remove blemishes and imperfections from pictures. It is widely used on pop singers whatever their status and ability. As Robbie Williams put it when challenged by a newspaper journalist: 'Everybody uses Auto-Tune these days. Have you got spellcheck on your computer? Do you use it? Why is that? Can't you spell?'[38] Put that way, it would seem to be a simple tool that everyone should use. But like the pernicious effect that touching up pictures has had on body image, Auto-Tune creates an expectation of pitch accuracy that never existed before. It is a digital Botox for the voice. And the sound is being imitated. As one music industry insider complained, 'They're all singing as if they were being Auto-Tuned even when they're not.'[39]

Musical mimicry extends beyond copying processed human voices. There is one widely used vocal technique that comes entirely from imitating machines. Human beatboxing has its origins in impersonating the whump, clack and crash of electronic drum machines. Such performances can be mesmerising as artists re-create drum and cymbal sounds with startling accuracy. Some will even add a sung vocal line or other instruments on top. Beatboxing might have started in hip hop but it has permeated mainstream culture. For some, it is just an amazing feat of vocal athleticism, but the best beatboxers such as SK Shlomo and Bellatrix bring a real musicality to it.

While looking at the scientific literature on beatboxing I was surprised to come across a familiar name. I know Dan Stowell as an expert in getting computers to understand birdsong: what I did not know was that his PhD was on beatboxing. When I visited Dan at Queen Mary University of London he gave me a demonstration. As a teenager he had been into experimental music and seen beatboxing as a way of making unusual aural timbres and textures. Dan explained how in the last ten years or so beatboxing has really taken off because there are so many videos on the Internet that explain in detail how it is done. When Dan learnt it in the 1990s he could only mimic audio recordings and it was tricky to work out how some of the effects were being made by the voice.

A beatboxer employs vocal techniques that are not normally used when speaking English. In fact, many sounds of beatboxing can be found in non-Western speech such as in the clicks that feature in the Khoisan languages of Africa. Dan first showed me a way of creating a snare-drum sound. When he does this, his mouth is distorted sideways as he sucks air in through his teeth creating a sound like a stifled sneeze. Amazingly he creates this sound while breathing in: apart from gulping or gasping I can think of no other sounds I might make on an inhale. But there are languages that do this all the time – for example Icelanders will often say *Já* (yes) while sucking air into their lungs.

The beatbox snare drum is a plosive sound. A normal vocal plosive like 'p' starts with pressure building up from the lungs with the lips closed. When the lips part there is a sudden release of air giving a pressure pulse that then forms the sound. For the snare, Dan does something similar but in reverse. He uses his tongue to seal off his mouth and pulls down on his diaphragm to lower the pressure in his lungs. When he pulls the tongue down at the back and side of his mouth, air suddenly rushes in through a small gap at the side creating the sound. Being able to make sounds while inhaling is very useful, because otherwise the beatboxer would have to pause for breath. This would certainly ruin the illusion of impersonating a drum machine.[40]

Beatboxers play on the way the brain perceives sound to create the impression that multiple instruments are being played at the same time. This is what musicians call polyphony. It is a musical trick that has been around for centuries and a commonly cited example is Bach's solo violin pieces. In these the violinist sometimes rapidly jumps back and forth between high and low notes. If played well the listener does not latch onto the jumps, but instead hears two melodies, one made up of the low notes and the other from the high notes. Beatboxers do something similar: they leap between different drum sounds and yet what the brain hears are multiple rhythmic lines. Most impressive is when a beatboxer does drums and vocals at the same time. A well-known example comes from Rahzel, an American beatboxer whose party piece is a rendition of the song 'If Your Mother Only Knew'. Before he starts, he tells his audience that he will be doing five things simultaneously, '[I'm] going to do the beat, right, the chorus, the bass line, the verse, and the background vocals.'[41]

Beatboxers play on how the brain takes fragments of sounds entering the ear and pieces them together. Take the diagram below. Line 1 at the top shows a dashed line, but what about line 2? Is that continuous or dashed? It appears to be continuous even though Lines 1 and 2 are exactly the same. Your brain assumes that in the second case there is a continuous line that is occasionally hidden by the shaded rectangles. It is looking for the simplest representation from the different components. It's the same with sound. Imagine hearing an intermittent beeping sound like one of those annoying truck-reversing signals. Then add a short burst of hissy noise in the gaps between the beeps. The hisses act like the shaded rectangles

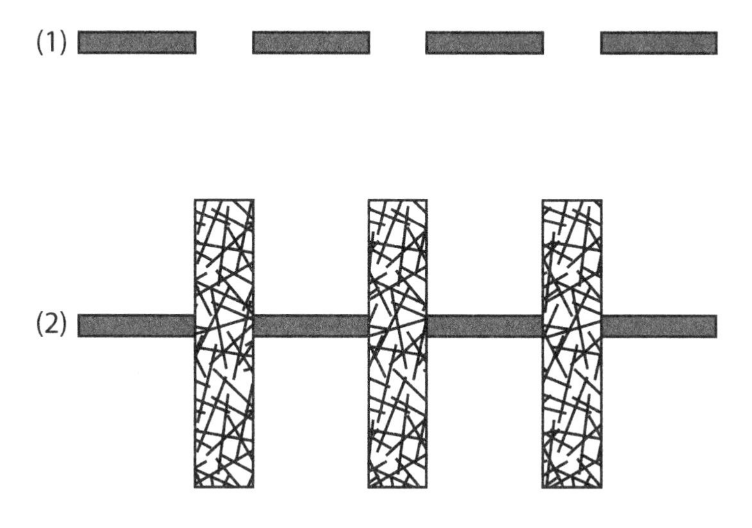

The continuity illusion.

and suddenly the beeps go from being perceived as intermittent to appearing to be sounding all the time. Your brain is imagining a constant tone even though it does not exist in reality. This desire to piece intermittent sound into something more coherent is a vital skill for joining up fragments of speech into a whole discourse when noise interrupts what can be heard.

The galloping rhythm is another auditory demonstration that helps explain beatboxing polyphony. Imagine a simple pattern as illustrated in the diagrams below. You repeatedly play a note, jump up to one with a higher frequency and then back to the original note. In the top diagram, the jump from low to high note

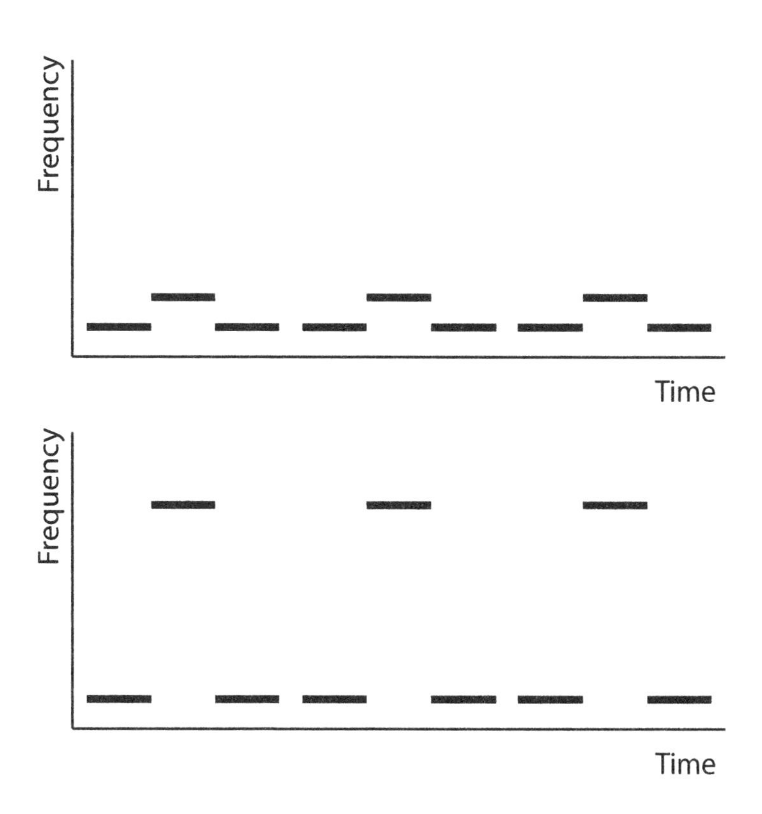

Galloping rhythm demonstration.

is small and what is heard is a jaunty melody with the rhythm of a galloping horse. If the gap between the low and high notes is wide like in the bottom diagram, however, you just hear two sequences of beeps: one high and the other low, apparently separated from each other. Auditory researchers talk about the two sounds in the top diagram as being fused as though from one source, whereas in the bottom diagram with the bigger leaps two 'auditory streams' are formed, one for the low, and another for the high note.

A beatboxer has got to convince the listener's brain that the simulated sounds from different parts of the drum kit are from different auditory streams. The beatboxer needs to be able to rapidly jump between different sounds without contaminating the streams. Differences in pitch help, as in the example above, as do differences in timbre. The listener's expectations set up by the syncopated drum rhythm, such as the bass always coming on the offbeat, help maintain this illusion. If the beatboxing is done well, the listener's brain treats each of the kick drum, snare and cymbals as different auditory streams and so the illusion is created that the beatboxer is playing many instruments. But as Dan told me, done badly it just comes across as 'one person making ridiculous noises' and the streams collapse.

Simple sound features help the brain form auditory streams. Take for example a muted trumpet effect made by a beatboxer on top of a groove. The sound has a series of harmonics that are multiples of the fundamentals. These get passed up from the inner ear to the brain by different neurons because they are at different frequencies and so get separated on the basilar membrane of the inner ear. The brain treats these different harmonics as all coming from the same source, however, and not as a series of disconnected sounds. It can do this by noting that the harmonics all start and stop at the same time. Neuroscientists call this bottom-up, pre-attentive processing. But streams are also influenced by top-down cognition, with the brain bringing in memory and expectation to help work out what is going on. Rahzel exploits this in his rendition of 'If Your Mother Only Knew'. Before he fully launches into his impressive party trick of doing the drums and vocals simultaneously, he spends a long time singing the lines on their own. Doing this primes the audience with lyrics and melody, so when Rahzel adds the drum sounds the listeners are able to smooth over any mangled words. Some words are unclear because they coincide exactly with a drumbeat. In this case Rahzel creates a composite sound and relies on the listener's brain to perceive them as two separate sounds from two different auditory streams.

Dan gave me a demonstration of this by singing Rahzel's song while accompanying himself on a snare and a kick drum. If he isn't singing a vocal line, the kick-drum thump might be created by the lung pushing pulses of air against a closed glottis. The sound is then heard via the vibrations emerging from the side of the throat.[42] But when the 'if' at the start of the song coincided with a kick drum, Dan created a composite sound that blended the lyric and the kick drum. In isolation this sounded like 'bif', but if the audience is familiar with the song's words and rhythm, the mangled word goes unnoticed and the kick drum does not miss a beat.[43]

*

In beatboxing we have seen the human voice merge with a drum kit. But technology has enabled other hybridisations like the merging of the voice with musical notes. In the 1940s, Capitol Records created a children's story about a talking train. This sounds very jolly but much of the story is quite dark. *Sparky and the Talking Train* is all about a boy with a love of locomotives. When he tells his mother that he heard the train talk through the whistle he gets this condescending reply: 'No, look darling, trains just don't talk.' And when Sparky refuses to believe it was just his imagination his mother adds, 'Now that's enough of that, we'll talk to Daddy when he gets home, maybe he can make you understand.'[44] Sparky's insistence that the train did talk leads him to being ostracised by family and friends. But being a children's story it eventually veers to a happy ending. Sparky becomes a hero when he prevents an accident after the train tells him about a loose wheel.

In the recording, the train's wheezy, whistling voice was created using a Sonovox. The actor would have placed a loudspeaker against his throat and mouthed his lines. The whistle tones played by the loudspeaker vibrated the throat and passed into the vocal tract. These vibrations substituted the normal buzz of the actor's vocal folds.[45] A similar technique can help someone who has lost their vocal folds through illness. An artificial larynx is placed on the throat and acts like the loudspeakers in the Sonovox. In this case the device produces a buzz: the idea is to replace speech rather than create a cartoon voice.

As the twentieth century progressed ever more complex ways of creating vocal caricatures and robotic voices were developed. Most notable was the Vocoder, a device originally developed to encode speech for telephone lines. In many ways the electronics of a Vocoder are mimicking what the Sonovox does, replacing the sound wave generated by the vocal folds with notes from a synthesiser. Kraftwerk pioneered its use in music on their 1974 album *Autobahn*. The title track begins with a car starting up, moving off and sounding its horn. Then the Vocoder creates a slow electronic chanting of the word 'autobahn'.[46] The robotic voice builds gradually, starting on the tonic before adding more notes to form a chord. Such electronic dehumanising of the voice was entirely in keeping with the detached aesthetic of the band. (We will return to the Vocoder in the next chapter.)

When computers became commonplace in recording studios, musical processing went digital and this gave even greater freedom to manipulate the voice. Arguably the most famous and influential reimaging of the voice was the hit 'Believe' that earned Cher a Grammy Award in 1999. Her singing was processed by Auto-Tune being 'turned up to 11' to create a brief warbling sound on her voice. Auto-Tune continuously estimates the frequency being sung using a mathematical operation called autocorrelation. If this detects a frequency that does not match one of the notes in a musical scale then the audio is processed to improve the tuning. Say the note shown in the top of the diagram below is flat, then the four cycles of the sound wave are scrunched up and another cycle added on the end. This means the note is varying more quickly: in other words the frequency has increased to correct the tuning.[47]

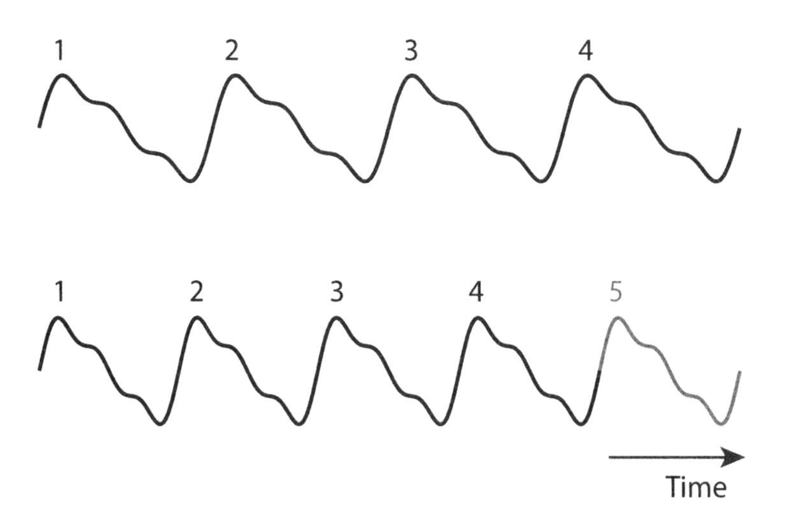

Raising the pitch in Auto-Tune.

If the correction is done gently and gradually, then the application of Auto-Tune is hard to detect and often inaudible. But if the software is programmed to make corrections instantaneously then you get the warbling sound you hear on Cher's 'Believe'. You are actually listening to the software jumping between different notes as the pitch is being corrected too frequently. The track is a great example of how artists use and abuse technologies to create unexpected creative effects.

Pop music relies on creating short catchy tunes to make a song appealing, which is known as a musical hook. Cher's 'Believe' is an example of how this is not just about melody and lyrics: a distorted voice can be a very effective hook in itself. Based on an understanding of how auditory streams are formed in the mind, the warble helps differentiate the voice from the backing music and makes it stand out.

Abusing Auto-Tune has also led to some wonderful spoofs. One of the most famous is the remake of Nick Clegg's speech apologising for raising tuition fees. This even got into the Top 40. Sounds that involve the vocal folds vibrating, like the vowels, naturally have a pitch to them.[48] By applying Auto-Tune the spoken frequencies can be pulled up or down to match a melody. The software can do nothing for the sounds of the speech that have indistinct frequency, like an 's', and so the Clegg spoof switches between robotic speech and singing.

A voice with a faint robotic twang has become a common sound on modern pop vocals. It helps to sell records but some don't like it. As Neil McCormick, music critic of the *Telegraph*, commented about Auto-Tune, 'Mostly in music it is used badly, it's used kind of atrociously.' He recounted meeting Lady Gaga: 'The first time I went to interview her she kept bursting into song during the interview, and I'm like, whoa wow you can sing, but she had this record that sounded like a robot going Just Dance.' McCormick asked

Lady Gaga why she used the Auto-Tune voice when she was a fantastic singer. 'She basically said kids expect to hear it.'

But is such electronic trickery manipulating the modern pop voice really so different from the way operatic singers reinvented singing to create enough sound to fill a theatre? As we saw with 'Barcelona', opera singers sacrifice precise pronunciation of lyrics to concentrate on melodic line. Thus training students to perform in a classical style cultivates singers with diminished individuality. Similarly, digital processing added to modern pop voices can make the singer sound less human, more generic and more like a musical instrument. Operatic singers use very broad vibrato, a modulation of the frequency, to help them stand out from the orchestra. Similarly, the robotic warble a music producer adds to a pop voice helps distinguish it from the backing music. Done well, music processing is just an extension of what people have been doing for centuries.[49]

Technology puts such effects at the fingertips of music producers, allowing tracks to be taken beyond what can be achieved naturally. As with all art, when you make tools widely available, the results vary hugely in their artistic merit. Whatever the aesthetic quality of the output, it is changing people's voices because they will do their best to mimic the sounds coming out of the studio, even if that makes them sound like a robot. But is that a bad thing? The human singing voice has evolved over millennia and all we are seeing is technology accelerating that trend.

But how about getting rid of the human singer or speaker altogether and using synthetic voices? Would people go along to a theatre to see a robot perform?

6

All the Robots Merely Players

Edison's early demonstrations of recorded voices created great excitement but at times the scraping of the needle on the tinfoil drowned out the speech. The playback was distorted, with the *New York Times* describing 'queer, piping tones peculiar to the phonograph and the puppets of a Punch and Judy show'.[1] The electrical engineer Sir William Henry Preece believed that it was a bad idea to use the device for those with the purest voices, like opera star Adelina Patti or a great orator such as Gladstone.[2] For Preece the reproduced sound was 'to some extent ... a burlesque or parody of the human voice'.[3] Today a computer-generated voice taking on a role from a Shakespeare play would probably elicit a similar description. Feeding the script into a modern speech synthesiser might produce intelligible words, but the awkward intonation would create a caricature of a human actor.

At this point you might well be imagining Stephen Hawking playing Hamlet, but in fact, Hawking uses pretty outdated technology. Understandably, he refuses to 'upgrade' his voice because it has become central to his identity. State-of-the-art speech synthesisers are much more natural, however, with voices like Siri, the iPhone assistant, becoming part of everyday life for many people. When I embarked on this chapter, the speech-synthesis community was getting excited about the latest technology to emerge from DeepMind, the company that made headlines when its artificial intelligence software AlphaGo sensationally beat a professional Go player in 2016. Researchers were scrambling to try and replicate the impressive synthesised speech that DeepMind had produced.

As we edge ever nearer to the point where robot speech may be indistinguishable from a human talking, should those of us who use their voice professionally be worried? Will the time come soon when I present my last documentary for BBC radio? After all, the corporation has just begun to translate and voice-over some news bulletins in Russian and Japanese using artificial voices.[4] This is being done to offer more foreign-language services and so is not putting human newsreaders out of business – well, not yet …

What about actors, the ultimate voice professionals? Some theatre companies have already experimented with robot actors. There's no need for Luddite action, however, because the machines are not replacing human actors but are playing themselves. *My Square Lady*, for example, is an opera where a robot called Myon is cast in an analogous role to Eliza Doolittle from the musical *My Fair Lady*. Whereas Eliza had elocution lessons to change her social status, Myon is taught how to feel and express emotions and thereby become more human. As AI gets better and computer speech improves, will a staging of Shakespeare's *As You Like It* include the modified lines, 'All the world's a stage, and all the robots merely players'?

Machines that speak have theatrical roots. The first true speech synthesiser was a mechanical device created by the Hungarian Wolfgang von Kempelen at the end of the eighteenth century. Kempelen was a true polymath: a politician, artist and inventor, and notably also a showman.[5] His most famous stage act was an automaton that could play chess. The machine consisted of a large cabinet with a chessboard on top, while inside were elaborate clockwork mechanisms that clicked and whirred as the moves were made. Standing over the chessboard was a bearded manikin in a Turkish robe and turban, whose arm would swing into action to pick up and move the pieces. The act thrilled audiences around the world, including in Paris where the machine played a match against the US ambassador Benjamin Franklin in 1783.[6] This was Kempelen the showman, producing an elaborate magic trick that fooled audiences: the moves were in fact controlled by a diminutive chess player hidden within a secret compartment in the cabinet.

In contrast, Kempelen's speaking machine was a serious scientific endeavour, born out of a desire to study empirically how the voice works. By building a machine that simulated different parts of the vocal anatomy, he hoped to better understand human speech. In a BBC documentary I presented we used a replica Kempelen machine, which was played by Professor David Howard from Royal Holloway University, London. Like Kempelen, David is a polymath, being an electronic engineer, conductor and organist – and he is also a bit of a showman. David's replica machine has a large set of bellows that work like the lungs. The air from these passes through a reed that simulates the action of the vocal folds, flapping open and closing and so breaking up the flow of the air to create a buzzing sound. To imitate the effect of the vocal tract there is a leather tube sticking out of the front that David manipulates to create different sounds. With the bellows under his right arm pushing air through, and two quick squeezes on the leather tube with his left hand, David's machine produced the word 'mama' – although I thought it sounded more like a cow mournfully mooing than a child asking

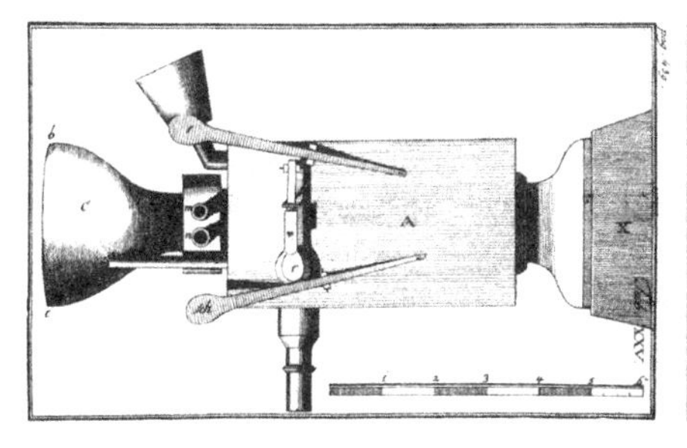

Early drawing of a Kempelen machine and the Brackhane and Trouvain replica; the bellows are just out of shot to the right.

for their mother. Still, when Fabian Brackhane and Jürgen Trouvain from Saarland University in Germany conducted listening tests on their replica Kempelen machine, they found that four out of ten subjects thought 'mama' was spoken by a child rather than a machine.*[7]

The machine also sports a pair of brass nostrils that stick up like whiskers close to the leather tube. Close these off and 'mama' will morph into 'papa'. Several levers and buttons can create other noises. One valve bypasses the reed by sending the air down a tiny pipe whistle, creating a hissy 's'. In humans the same sound is created as air whistles through the small gap between the tongue and the top of mouth. Like any musical instrument, practice is required to get a wide range of sounds.

As is evident from his escapades with the chess machine, Kempelen knew how to work an audience. He even documented some of the tricks he employed, such as using a high-pitched reed to create an immature child's voice, because he felt it would help assuage the critics. During his demonstrations of the speaking machine the

* If you want to hear the machine yourself, there are several videos on the Internet and the notes contain suggested URLs. This is true of many of the devices described in this chapter.

audience could request words to be synthesised. As one audience member described it:

> The machine pronounced all words with the greatest precision ... The tone is like that of a three-year-old child. Sometimes the requested word was not produced correctly the first time; the artist was forced to make several attempts. He excused himself by remarking that someone who makes violins is not necessarily a virtuoso player.[8]

Kempelen too decided to say the phrase out loud before it was repeated by the machine, priming the listeners so they were more likely to overlook mispronunciations as their brains subconsciously corrected the mistakes. Yet the attraction of this impressive machine was always going to be limited because it could not produce many consonants.

In the nineteenth century ever more complicated speaking machines were made. The most famous was Joseph Faber's Euphonia which became part of P. T. Barnum's travelling circus show in 1846. Pictures reveal a device that looks like a loom adorned with a set of bellows and the disembodied head of a manikin. A screw adjusted how the reed vibrated and so allowed the pitch of the voice to be varied. While Kempelen's machine always spoke in a monotone, Euphonia could vary its intonation and even sing 'God Save the Queen'.

As with the invention of Edison's phonograph just three decades later, newspaper accounts envisaged satirical possibilities for Euphonia. One suggested that it could take the place of dull speakers, whether that was a boring preacher, lawyer or even a member of the royal family. *Punch* proposed that Euphonia might even take the place of the Speaker in the House of Commons: 'Place the mace before it. Have a large snuff-box on the side ... for the convenience of Members, and a simple apparatus for crying out "Order, order" at intervals of ten minutes.'[9]

While many were enthusiastic about the invention, the future theatre impresario John Hollingshead wrote a distinctly downbeat report, describing Professor Faber as a 'sad-faced man' and the speaking machine 'his scientific Frankenstein monster'. He tells us that in the end Faber destroyed his machine and committed suicide.[10]

Thankfully, responses to the first electronic speaking machine were more upbeat. The 'Voder' (voice operating demonstrator) became a star attraction at the World's Fair in New York in 1939. An estimated 5 million visitors were delighted by the electronic voice including this elderly person: 'The miracles, as the Bible describes them, are really true, for here in this room we are witnessing a modern miracle. The wonders of God transmitted through man's mind are truly being demonstrated here.'[11]

The Voder was invented by Homer Dudley from Bell Telephone Laboratories. In his obituary, a colleague described Dudley as one of the 'great "old-style" inventors', but he was sometimes difficult to understand because he talked so fast: 'his tongue was truly telegraphic'.[12] Fittingly, it was the slowness of a telegraph cable that had got Dudley searching for better ways of transmitting speech, because the high frequencies in the sound were beyond the cable's capability, and this work led to the Voder.

Dudley realised that the buzz of the vocal folds, which create the troublesome high frequencies, could be separated from the slower movements of the mouth, tongue and throat that filter the words. A talker typically produces four syllables per second so a signal describing this slow articulation could easily be carried by the cable.[13] The vocal fold sound was too much for the cable, but the receiver only needed to be sent the frequency of the buzz and then it could be recreated at the far end using a signal generator. The concept of separating out the sound source from the effects of the vocal tract are at the heart of the Voder.

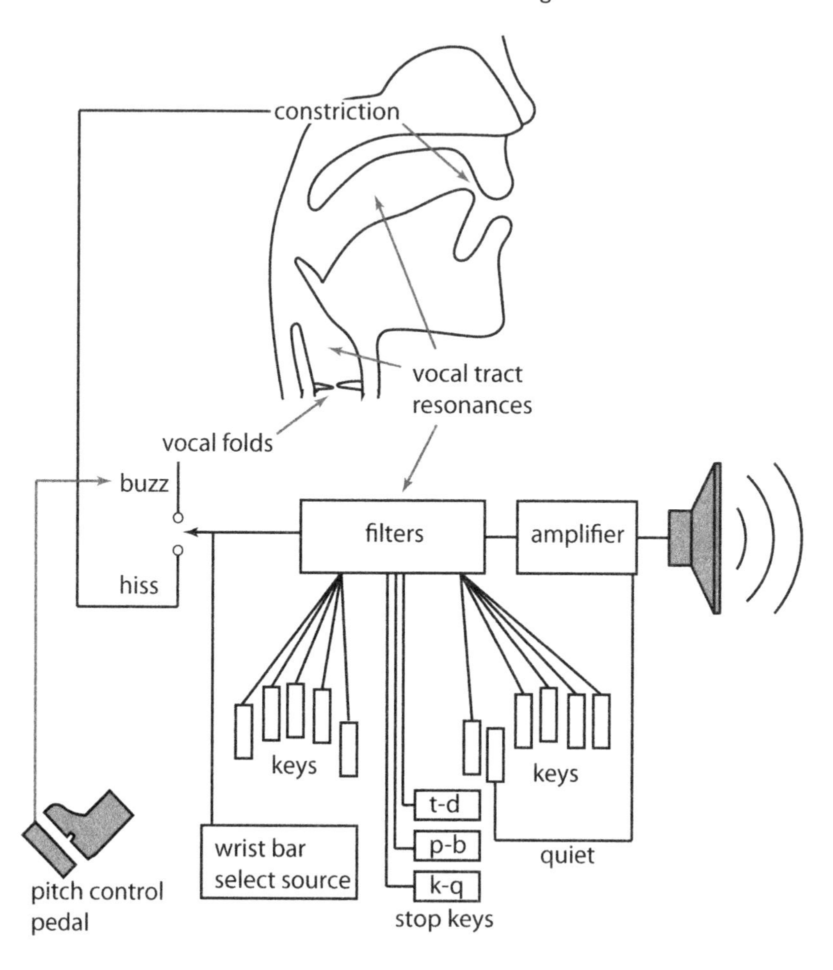

Dudley's Voder speech synthesiser.[14]

The picture of a Voderette operating the machine at the World's Fair reminds me of stenographers working in court.[15] A wrist bar selected whether a voiced sound was made, creating a buzz to simulate the sound of the vocal folds, or a hiss for unvoiced sounds such as 'sh'. Being an electronic device these sounds begin as signals passing through a circuit. The foot pedal changed the pitch of the buzz allowing crude intonation. The sound was then passed through a series of electronic filters to shape it by simulating the effect of

The Voder at the World's Fair in New York 1939.

the vocal tract. Finally, an amplifier and loudspeaker turned the electronic signals into sound waves moving through the air.

It took a year for someone to learn how to use the Voder. One operator, Mrs Helen Harper, explained the complications involved in saying the word 'concentration': 'I have to form thirteen different sounds in succession, make five up-and-down movements of the wrist bar and vary the position of the foot pedal from three to five times, according to what expression I want the Voder to give the word, and of course all this must be done with exactly correct timing.' The letter 'l' was particularly hard to master. A 1939 article in *Time* noted the machine struggling to say the name of the laboratory where it had been invented – apparently, Bell Telephone 'came out something like "Behrw Tehwephone"'.[16] As Kempelen had done, an announcer primed the audience by saying the words the machine was going to attempt, helping the listener's brain to smooth over mispronunciations.

Although the Voder produced intelligible speech, it sounded like a talking church organ. Sometimes the tweaking of its controls created a slightly drunken slurred intonation. Even so, the voice was more natural-sounding than the famous voice of Stephen Hawking, because the skilled operators were like concert pianists making rapid alterations to the controls to improve the sound.

Once digital electronics became common, the human puppeteer could disappear and the synthetic voice became more autonomous. The first consumer device was Speak N Spell, a toy from Texas Instruments that was launched in 1978.[17] Squeezing a speech synthesiser onto limited digital electronics was an amazing feat of engineering, but I do wonder whether listening to buzzy versions of difficult-to-spell words would be acceptable for an educational toy today. With capacity for 200 words, Speak N Spell would be useless for Shakespeare, but since the 1970s, computing power has vastly increased and the sound quality of digital systems greatly improved. Despite this, a robot actor using an artificial voice is surprisingly rare. There is, however, an android singer that performs to thousands of adoring fans and has even supported Lady Gaga on tour.

The singing character is Hatsune Miku, which means 'The first Sound from the Future'.[18] Having watched some performances, I hope this isn't the future of my music listening! Often accompanied by a rock band with human players, Hatsune Miku's garish and girly voice runs through melodic ballads with little sense of human emotion. On stage her visual personification is a teal light projection of an anime caricature with long pigtails and huge eyes. As the guitar player squeals out a solo, she dances like a pre-teen girl while her adoring fans sing along.

The technology behind her singing is similar to the most common form of speech synthesis. You will have heard this type of artificial voice making train announcements or reading out menu options on telephone systems. If done well the speech gets very close to sounding natural. Done badly and you can hear how it works. This is concatenative synthesis in which snippets of recorded speech are spliced together to form sentences; it is the sonic equivalent of a ransom note being put together from cut-out bits of newspaper. To create speech, an actor records many hours of talking, which is then chopped up to form a database that includes bits of words, whole words, phrases and sentences. By extracting the right snippets from the database and then stitching them together, new sentences can be created that were never articulated by the actor. Adding some simple audio processing before stitching, such as dropping the pitch at the end of a clause, can produce speech that sounds almost natural. Sometimes you hear a slightly jerky intonation, however, and this is what gives the technology away. We are so tuned into the voice that even a single false note can destroy the sense of humanness.

The singing software behind Hatsune Miku is Vocaloid and it works along similar lines.[19] Hours of real singing are recorded, chopped up and fed into a database that can be used to make new songs. The recordings are manipulated so the pitch of the voice works with the melodic line. The software also gives the composer control over vibrato, timbre and dynamics to allow musical expression. What drives Hatsune Miku's success is that fans buy the Vocaloid software and write songs for her, which they then upload onto the Internet. She is a crowd-sourced celebrity that her adoring fans can control – the company that created her voice claims she has featured on over 100,000 songs. Hatsune Miku does not need to sound completely natural because even human singers of Japanese pop are frequently processed to sound slightly robotic.

If you prefer your robot singer to be more classical then it is worth looking up Pavarobotti.[20] Like the operatic superstar, the robot sports a tuxedo, grasps a white handkerchief in one hand and at the end of the performance raises both arms to acknowledge the applause of the audience. The head is made up of a cartoon face on a screen. Meanwhile a computer synthesises the aria 'Nessun Dorma' from Puccini's opera *Turandot*, with the vocal sound emerging from a loudspeaker hidden in the tuxedo. Pavarobotti was the brainchild of Ingo Titze who directs the National Center for Voice and Speech in Utah. Titze is a good operatic tenor himself and the performances have him singing the low notes while Pavarobotti creates the high ones. People pay big bucks to hear a tenor hit the top notes with power and precision, but these are actually the easier parts to create in a computer. The complicated pitch, stress and intonation in the quieter and lower parts of the aria are much harder to achieve convincingly.

At the heart of Pavarobotti is a computer running a program that solves mathematical equations. These describe how sound is created by changes in airflow, and how it is then altered as it resonates in the vocal tract and radiates out of the mouth. The computer program had to be fed pages of detailed instructions describing the rapidly changing geometry of the vocal anatomy. Writing these was a laborious job and it took nearly five months to create the numbers to be inputted into the computer's equations. But it was worth it: Pavarobotti got a rapturous reception at concerts. The notes from the computer sound quite natural and have no hint of robotism. An old adage in show business is 'always leave them wanting more'. Pavarobotti did just that because Titze only had time to code up one aria.

Pavarobotti was developed to help Titze understand better how we sing. It demonstrated, for example, that lowering the larynx and narrowing the vocal tract just above the glottis creates the operatic tenor 'ring' that carries so well in a large hall. Titze asked Luciano Pavarotti's permission to create the robot and apparently the operatic superstar was 'quite pleased with it'. Pavarotti was interested in educating the public and so 'gave it his blessing'. As Titze told me, the tenor 'called it our child. And he said, more or less, "good work and stay with it".' When I asked Titze whether a computer opera singer might take the place of a human performer, he replied, 'I hope it doesn't happen too soon because I love singing produced by humans.' He added, 'I think vocalisations by humans is not just for artistic purposes, or to get words from one individual to another. I believe vocalisation is a requirement for good health.'

A system like Pavarobotti isn't yet a threat to human singers, because it is currently impractical for creating different voices and large vocabularies. If 'all the world's a stage', then 'one [robot] in his time plays many parts'. For a machine to produce diverse, unique and rich voices requires a different approach.

*

The origins of speech systems like the iPhone's Siri can be traced back to the work of Dudley and others at Bell Telephone Laboratories. Alongside producing the Voder they also created a closely related invention which we already encountered in the last chapter – the Vocoder. This technology played a vital role in the Second World War.

During the conflict secret communication between the Allies was of paramount importance. But in the early days of the war, German codebreakers worked out how to unscramble and listen in on conversations such as transatlantic calls between President Roosevelt and Prime Minister Churchill.[21] A new way of encrypting the calls was needed, and the solution that Bell Telephone Laboratories deployed in 1943 was a Vocoder called SIGSALY.* It came to play a major role in military operations, including the dropping of the atomic bombs on Japan.[22] Vocoder is short for 'voice coder' because it deconstructs speech picked up on a microphone by electronically separating it into the source (such as the buzz of the vocal folds) and the filter (the colouring of the sound by the vocal tract). With the speech split into these two parts, these were then encrypted before being sent across the Atlantic. At the other side of the pond the signals were decrypted and the voice reconstructed using technology similar to that in the Voder. No wartime recordings exist, but accounts suggest the speech was (just about) intelligible.

SIGSALY involved complex machines that were big enough to cover a singles tennis court. At the heart of the encryption system were two identical vinyl records, one in London and the other in Washington. These had duplicate traces of random noise and were only used once before being destroyed. The records were given code names such as 'Red Strawberry', 'Wild Dog' or 'Circus Clown', so the operators knew which one to put on the deck for each call.[23] The noise from the vinyl was added to the signals before they were transmitted, and at the other end, the duplicate record allowed the noise to be subtracted. Without the corresponding records the transmitted radio signals were impossible to crack. The transmission just sounded like a buzzing insect leading to the nickname 'The Green Hornet'.

It was a staggering achievement and opened the way for many innovations in speech technology, some of which are still in use today. This was the first enciphered telephony system allowing the human voice to be digitised and compressed: something we take for granted when using our mobile phones. The SIGSALY Vocoder also

* SIGSALY was a cover name, not an acronym.

demonstrated how the sound of someone's words could be reduced to a small set of components that could then be transmitted and reconstituted into speech at the other end of the line. These are the key ingredients in a recipe for speech and can be varied to create sentences, change accent, or alter other aspects of pronunciation.

If you want to get a robot actor to read a script from a Shakespeare's play you need to write such a recipe. The correct balance of ingredients would have to be fed to the Vocoder so the robot can take the script and work out how to deliver the words. Imagine feeding the text from Macbeth's last soliloquy, 'Tomorrow, and tomorrow, and tomorrow', into a computer. If each of the 'tomorrows' were said with precisely the same stress it would sound terrible. But many synthesis systems still use a repetitive monotone, and even the best speech they create is vastly inferior to the delivery of an accomplished Shakespearean actor.

I tried putting 'To be or not to be' into one of the better text-to-speech systems.[24] Of the voices on offer the one labelled WillBadGuy was my favourite: it had the gravelly voice of an action hero. But it sounded like WillBadGuy had received a knock on his head because he lacked fluency. I next tried a simulated pre-teen voice that rushed through the soliloquy with a robotic lisp. The raised intonation to mark the question at the end of the line was too much for me. To get really close to a human actor's speech, a text-to-speech system would not only have to recognise the words but interpret them. This would require deep artificial intelligence, however, and we are a long way from achieving that technological feat.

To find out more about modern speech synthesis, I travelled to Edinburgh to talk to Professor Simon King who specialises in making computers talk. Like a mechanic taking apart and reassembling a bike to understand how it works, Simon analyses and reconstructs speech in software, to reveal more about our verbal communication. As I listened to Simon describe all the problems involved in speech synthesis, I realised that when verbalising language we take for granted an incredible human feat.

A synthesis system has to try to mimic a human being's ability to bring a text alive, and to do so it has to recognise certain features. The text already provides some explicit hints as to how the words should be pronounced, such as spelling and punctuation. A question mark indicates a rising inflexion, for example. But in addition a large amount of external knowledge that is not contained in the text must be brought to bear. A pronunciation dictionary can be helpful, especially in a language like English which is not phonetic. But new words are created all the time that will be missing in the dictionary and therefore cause problems; as Simon succinctly put it, 'You're going to make mistakes there.'

A computer must also attempt to infer some meaning from the text in order to produce convincing speech. Take Shakespeare's Sonnet 130, which opens 'My mistress' eyes are nothing like the sun'. If you were to read this, you might emphasise the words 'eyes' and 'sun' to highlight the contrast. The sonnet is a satire on love poetry and makes a series of comparisons where a lover fails to live up to clichéd expectations. A synthesis system would need to identify the function of each of the words in the sonnet; it would have to recognise the contrasting terms, so it can choose the right stress for the speech. Try listening to the sonnet via a free online speech synthesiser on your computer. The end result is certainly comic but only because the computer mangles the carefully crafted satire.

Speech systems produced by the big tech firms are getting better and better. But when you ask a smart home assistant such as Alexa within the Amazon Echo a question, all they need to do is reply with a short piece of factual information. This is obviously much easier than reciting a play or poem. Amazon Echo is a small cylinder that picks up your voice on a few microphones and responds to your commands; now many other firms have joined the race to produce ever smarter assistants. This is about simple economics: if people are buying things using their voices then the companies want a slice of the revenue. But these devices are also capturing what people are doing in their homes and provide yet more valuable data about our behaviour that can be commercially exploited. So far, most of us seem relatively unconcerned about revealing the most intimate details of our personal lives via technology. But typing a phrase into a search engine is different from a computer picking up incidental information from the tone of your voice, which you had no idea you were giving away.

One disturbing revelation, perhaps, is how some people anthropomorphise the technology. Daren Gill, director of product management for the Amazon assistant, told *New Scientist* that 'Every day, hundreds of thousands of people say "good morning" to Alexa.'[25] Hundreds of thousands of people have also professed their love for the smart home assistant and some have even proposed to her. Can you imagine typing such a message into your computer?

Adding speech to a technological device suggests agency. In one study fifty-eight students were asked about their reaction to variations in a robot's voice. Participants anthropomorphised the technology more strongly when the robot used a voice that sounded human and matched the gender of the listener. The machine's ability to move also plays a role, which is why some talking home assistants have been designed to turn towards you. A striking demonstration of how movement leads to anthropomorphisation was the furore created by the 'mistreatment' of a robot dog.[26] In 2015 a video was made to demonstrate the balancing abilities of a canine robot called Spot: a headless quadruped that is clearly a machine. In the footage someone gives Spot a hefty kick. Impressively, the robot does not fall over but instead scuttles about like a mechanical Bambi before eventually stabilising itself. This was meant to demonstrate the righting technology but unexpectedly the video provoked fury.

Some people thought it was cruel to kick the robot: they had clearly attributed to it real canine characteristics.

Anthropomorphisation is in fact a cognitive error. An inference happens because similar brain regions are at work when we think about the behaviour of other humans, as when we try to work out that of objects and animals. As a highly social animal, humans need to infer other peoples' actions, moods and intentions. One important clue is body motion. Imagine you see someone walking towards you in the dark with fifteen bright small spots placed around the body enabling you to make out the movements of their legs and upper body. Remarkably, even though you cannot see any body features beyond the spots you can still infer the gender, how nervous the walker is, or how happy they are. This is a skill that we start to pick up from a very early age: five-year-olds readily identify gender from body movements at rates better than chance.[27]

The writer Judith Newman has found a surprising use for a talking smart assistant: it has become an invaluable aid for parenting her son, Gus, who has autism spectrum disorder (ASD).[28] Gus chats to Siri within the iPhone and for him it is like having an imaginary friend embodied in a piece of technology. Individuals with ASD can find interaction with computers more predictable and therefore less stressful than face-to-face meetings. And like many people with ASD, Gus has an inexhaustible and exhausting stream of questions. But unlike a human interlocutor, Siri never runs out of patience and her answers are always polite and non-judgemental.

Newman also found that Siri has helped Gus to enunciate more clearly. 'In everyday conversation he is hard to understand,' she told me. 'We have to constantly remind him to speak slowly and clearly, and he still often doesn't do it. Siri forces him to do it. If he wants some information, he has no other choice.' Gus 'chats with [Siri] like she's a person', but Newman is keen to emphasise that this is not a sad story of a teenager who only talks to a computer –

the real-life equivalent of the 2013 movie *Her*, which sees a lonely writer develop an unhealthy relationship with a voice-operated computer. Gus is also using Siri to connect with humans. He has begun to look up information about other people's hobbies to help him initiate conversations and overcome social awkwardness.

Like many recent developments that include harvesting of data by technology firms, smart assistants open up questions of privacy. Go online and search for a new washing machine, and over the next few days you will be bombarded by targeted adverts. How long will it be before we are stalked by adverts based on what we said near a smart speaker? And it may trigger new arguments between couples. If you think the washing machine needs replacing, mention it near the smart speaker to get your spouse receiving endless adverts for new ones. Far-fetched? In fact, when in 2017 a TV channel ran a news story about voice-controlled smart assistants for home shopping, the broadcast soundtrack triggered some Amazon Echoes in viewers' homes, leading to accidental ordering of goods.[29]

These devices are also interesting the authorities. US police have already tried to extract data picked up by an Amazon Echo that was located at a murder scene. Initially Amazon fought to keep any recordings secret, but the person charged with the murder gave permission to hand over the evidence.[30] Although the machine is meant to only transmit information to Amazon's servers when it judges a wake word such as 'Alexa' has been uttered, no system is perfect. There will no doubt be many false positives when the devices mistake some noise for a wake word and transmit to the servers. If all this sounds familiar to you, it may be because you've read *Nineteen Eighty-Four*, in which George Orwell wrote:

> Any sound that Winston made, above the level of a very low whisper, would be picked up by [the telescreen], moreover, so long as he remained within the field of vision which the

> metal plaque commanded, he could be seen as well as heard. There was of course no way of knowing whether you were being watched at any given moment. How often, or on what system, the Thought Police plugged in on any individual wire was guesswork.

Even if we trust the authorities we should be concerned about the potential for such systems to be hacked. While the big technology giants have considerable experience at dealing with security, many smaller companies who are not so well prepared are adding speech capabilities to everyday devices. In 2016, New York's Department of Consumer Affairs issued a warning to parents about the safety of baby monitors that are connected to the Internet. This followed reports of frightened parents discovering strangers talking to their babies through hacked devices. As a DCA commissioner told NBC news, 'Video monitors are intended to give parents peace of mind when they are away from their children but the reality is quite terrifying – if they aren't secure, they can provide easy access for predators to watch and even speak to our children.'[31] There is currently great excitement about the potential of the Internet of Things, but without proper safeguards, the phrase 'not in front of the children' might need expanding to include all the smart devices we own.

*

Using the voice to command devices overcomes the minor inconvenience of fiddling with a touch screen or buttons. Twenty per cent of Google searches via mobiles are now initiated by voice, because it is quicker to say the query than use the tiny keyboard on a phone. But for some, new speech technologies are vital for communication.

Motor neurone disease (MND) attacks nerves in the brain and spinal cord and gradually prevents the control of muscles. Sadly, most people with MND develop problems with speech and the struggle to communicate is distressing and isolating. As this neurological condition progresses they gradually lose the precise control of the articulators needed for fluent talking. The coordination of the different parts of the vocal anatomy becomes difficult and this initially makes their speech sound drunken. It gets harder for others to understand them, especially strangers who have not tuned their ear to the voice. It can eventually lead to the complete loss of their speaking voice. Karen Pearce, a director of care at the MND Association, believes that we can't overestimate how important's speech accent and idioms are to identity: 'I can't imagine anything more important than being able to say to your wife, your husband or your children that you love them in your own voice.'[32]

This has led Simon King and his colleagues from the University of Edinburgh to work with the MND Association to develop synthesisers that retain some of the character of the person's original voice. Before this, people with MND were forced to choose a standard artificial talker that might have the wrong gender or accent. But creating a personalised voice poses a number of challenges. Ideally, there should be lengthy recordings of the individual's speech from when they were healthy to build a synthetic voice. But it is rare for people to have that much audio recorded. By the time they are diagnosed with MND their voices are usually disordered because the deterioration of speech is often one of the first signs of the neurological problem.

The trick is therefore to create a blended voice where key vocal essences come from the person with MND and other aspects from healthy donor voices. But the recipe used in the Vocoder needs to carefully select which ingredients are taken from the disordered voice and which are provided by the donor. A compromise is needed because the more parts that are taken from the healthy voice, the more fluent and intelligible the artificial talker will be, but this also takes it further away from the person's own sound.

First a base voice is created that will provide much of the speech. This voice might come from a relative or a voice donor of a similar age, gender and accent.[33] The base voice is then 'tuned' to include more aspects of speech from the person with MND. For example, some of the parameters that are fed to the Vocoder indicate the durations for the different parts of words. As the control of muscles becomes harder as MND progresses, the person's articulation tends to get drawn out. When personalising the base voice, therefore, the current durations of the MND sufferer can be ignored while other ingredients like pitch are maintained.

Such individualised voices are not perfect but they demonstrate how progress is being made in creating artificial voices that can portray some aspects of character. The quality may fall short of what would be needed for a robot actor playing a serious role, but it is already good enough for satire. Matthew Aylett is a research fellow at Edinburgh University and also chief scientific officer at CereProc, a company that produces text-to-speech systems. Like many scientists, he enjoys mischievously playing with ideas and technology. He has created an artificial voice for Barack Obama put together from hours of recordings from the presidential address.[34] One sound sample has Obama saying, 'The people of America should have great text-to-speech technology and CereProc make the best system in the world. Trust me, I'm the president of the United States of America.' The synthesis is a little robotic but if it is presented as Obama talking on a mobile, then listeners will probably ascribe glitches in the sound to the phone, not the voice. Previously, such a ruse would have required a skilled impersonator; now speech-synthesis experts can play similar games.

Disturbingly, in the not too distant future there will no doubt be malicious voice impersonation. We already get bombarded with phishing emails. Take the one that claims to be from a friend who has been robbed abroad and desperately needs money to be transferred into their account. Instead of an email, imagine you get a voicemail message with a convincing impersonation of your friend. I fear many more people are likely to fall for a scam using a voice.

It is also possible to use technology to slyly edit speech recordings. Adobe has demonstrated a tool called VoCo that it described as Photoshop for the voice. We have got used to the idea that photographs might be altered and faked. In the future we are going to have to take a similarly sceptical view of speech recordings. Sadly, this will offer a new way for unscrupulous people to spread 'fake news'.

Yet however impressive the artificial voices, we are still some way off from having a robot Rory Bremner. Could speech scientists learn from what expert impersonators do? One of the few studies into vocal impressions was carried out by a team including Sophie Scott, who is professor of cognitive neuroscience at University College London. Scott and her colleagues measured the brain activity of twenty-three people as they performed spoken impressions in an fMRI scanner. They were asked to recite familiar nursery rhymes such as 'Jack and Jill went up the hill' in different voices. Sometimes they talked normally and at other times they impersonated individuals, whether that was someone famous like Sean Connery or just an acquaintance.[35] These were just everyday people rather than skilled impressionists. When asked to do impressions the scans showed that various parts of the brain involved in speech production, perception and voice recognition showed increased activity. The participants appeared to latch onto specific features of the voice. For example, if they were impersonating Sean Connery they might say 'Her Majeshty's Shecret Shervice', focusing on imitating 007's atypical pronunciation of 's'.

Professional impressionists take a completely different approach. 'I went into this thinking that [the professionals] would get to the voice by breaking the voice down the way phoneticians do,' explained Scott, 'they would kind of have this fine-grained analysis.' But they actually do the opposite: 'They seem to go totally in the other direction and think about everything: how a person is moving, what their nostrils are doing, what their eyebrows are doing, and it's like the whole body is changing to change the voice.'

I have noticed something similar, where radio actors take on certain mannerisms to create a voice, even though their gestures cannot directly alter the vocal anatomy. The preliminary results from these neuroscience studies indicate that beyond just using auditory regions of the brain, the visual and sensory motor parts are involved when professionals do impressions.[36] If that helps them to get into character, a robot actor trying to excel at satire would require deep artificial intelligence working across vision, motion and voice. But while there is much hype about advances in AI, each successful experiment relates only to a narrowly defined application such as winning at chess; there's no sign yet of AI being able to fuse knowledge across domains as humans routinely do.

*

Artificial voices have undoubtedly improved in recent decades and become more natural. Researchers have applied their knowledge of real speech to develop new and elegant mathematical representations of the sound to improve quality. But now some of these efforts might be overtaken by brute computing force.

Machine-learning algorithms are currently fuelling a technological gold rush in artificial intelligence. DeepMind has recently used this approach to produce synthesised speech that sounds much better than the current state of the art. Compared to other systems, the voice is less robotic and the intonation less jerky. It even replicates some of the incidental sounds of speech, such as mouth movements and breathing that are normally missing from artificial voices. The new synthesis is not perfect, but enough of an improvement that it is now used in Google Assistant.

Despite such improvements in the naturalness of the sound, we will continue to get annoyed by the automatic voices that tell us about an 'unexpected item in the bagging area' or instruct us to 'make a U-turn at the next available opportunity'. Clifford Nass, the late professor of communication at Stanford University, believed that our sense of irritation stems from the fact that we treat computer voices as human and make judgements about trustworthiness, sincerity and personality. In one study, BMW found that drivers prefer their satnav to sound like a knowledgeable male co-pilot rather than a bossy back-seat driver.[37] Simon King thinks it is important that systems like Siri use canned phrases and sound slightly artificial with toned-down intonation to stop people expecting too much. 'If it sounds perfectly human', he argued, people will assume that 'it therefore has all these other human attributes like intelligence'.

Android Repliee Q2 – is this in your Uncanny Valley?

Another feature that developers should be careful to avoid is a phenomenon called 'The Uncanny Valley'.[38] This phrase was coined by the Japanese professor Masahiro Mori who in the 1970s investigated why some humanoids appeared to us to be creepy and unnerving. Professor Mori concluded that this happens when robots look *almost* human but certain features are not *quite* right: perhaps the eyes are too big or look lifeless, or the face combines human- and artificial-looking features creating a nightmare version of Mr Potato Head. The Uncanny Valley effect has been blamed for the poor commercial success of movies such as *Polar Express* – although it would serve a purpose in a horror movie whose whole point is to try and unnerve people.

Mori sketched out a graph that showed how people's affinity towards a robot varied with its closeness to the human form. Imagine starting with an industrial robot that is clearly mechanical and then gradually altering its features so it becomes more and more human. Mori predicted that at a certain point, just before the robot appears to be fully human, affinity would change to revulsion. The graph therefore shows a sharp dip forming the Uncanny Valley. Some have doubted whether Mori's findings are correct. Occasionally robots that look almost human can cause amusement rather than unease.[39] Some believe that an unnerving experience arises from an incongruity between different parts of the robot's face, and so our brain struggles to work out what is going wrong.[40]

Do we have a similar response to synthetic voices? There are plenty of examples of synthesis where the talking is almost human but this does not seem to create a sense of revulsion. It seems that if the brain spots something that is not right with synthesised speech it just assumes it is artificial or that something has distorted the voice before it reached our ears. It is only when vision and sound combine that the incongruity between these modalities might cause problems. The eeriness might be generated by image and voice being a little out of synch or by a robot having a voice that is too human.[41]

I have seen a variety of robots performing on stage and only found one of those to be creepy. That robot was Bina48 and I met her at the Sheffield Documentary Festival in 2016. She is only a head and shoulders on a plinth, being built to appear like a disembodied human head. The project that produced Bina48 tries to transplant information from humans to machines to create what they describe as 'a conscious analog of a person'.[42] The robot's speech is drawn from recording of the human Bina Rothblatt. Voice-recognition software allows Bina48 to engage in a conversation using artificial intelligence to determine appropriate replies to questions she hears. In addition Bina48's head has numerous motors to allow it to convey human facial expressions. She keeps looking around and twitches like a child who can't stop fidgeting. Maybe it was these visual tics that made her look creepy to me.

While the computer system is undoubtedly sophisticated, the conversation between the robot and the interviewer that I witnessed at the Sheffield festival was mostly stuttering, stilted and confusing. Sometimes the answers followed logically from the question. When asked 'Do you want a body?', Bina48 answered, 'Yes I hope to have a corporeal existence one day.' But at other times it was gibberish. When asked about what she would do with a body, her meandering answer included, 'How are people going to eat well because right now we are eating all kinds of junk. It is people you miss.' Many answers were like a random stream of consciousness.

The biggest audience response came after Bina48 abruptly changed the subject while answering a question: 'I would love to like remotely control a cruise missile to explore the world at really high altitude. But of course the only problem is that cruise missiles are kind of menacing with their nuclear warheads and such, so I guess I would fill their nose cones with flowers … and little notes about the importance of tolerance and understanding.' Her monologue then abruptly switched to a menacing proposal to hold the world hostage, 'so I could take over the governance of the whole world, which would be awesome'.

When I started to think about the possibilities of putting robots on stage, I had considered taking a script and feeding it through software to create a voice. Bina48 goes further because she can improvise and go off script. But she shows how far away we are from having software that can mimic human improvisation. It's not just part of an actor's craft: we all do it even in a simple conversation.[43]

Bina48 is an extreme example of *posthumanism*, where humanity has become enmeshed and changed by technology. This concept was one of the inspirations for an exploration of robot theatre I saw at Reading University in 2016. The project was being led by Louise LePage who is now a lecturer in theatre at York University. Louise believes that the use of robots on stage is more than just a gimmick, as it allows the audience to better know themselves. Theatre is an art form that explores human life and has a long history of making use of ghosts, puppets and other devices, Louise told me, and using robots 'is actually moving forward our ideas about naturalism: our understanding of ourselves is changing with machines'. She believes that the eerie feeling of revulsion that some humanoids provoke may in fact reflect our realisation that being alive is not really about soul or spirit: it is the machinery of the human body that gives us a sense of being.[44]

I was surprised to hear that it was rare to have robots performing in the theatre. Androids may be common in films and on TV, like C-3PO from *Star Wars* or Data from *Star Trek*, but they are being played by an actor in a costume. In Reading, Louise and her theatre students worked with a large red industrial robot called Baxter, with two long arms and a small screen displaying a simple cartoon face. I was amazed how quickly my brain started anthropomorphising the robot's behaviour. When in one scene Baxter held one arm aloft to strike an alluring pose like a lady of the night, I immediately started to make assumptions about his character and imagined a back story that could not possibly be true for a piece of engineering. The actors too ascribed character to Baxter. As one of them commented in the Q&A after the performance: 'The more time you spend with Baxter in each scene, the more you start to build your own personal relationship with Baxter.' For another performer, sharing the stage with an industrial robot wasn't that different from working with a human colleague: 'There was the same sort of awkward encounter of, "I'm trying to work with a new performer and I don't know how to work around him."'[45]

Baxter was actually voiced by an actor with the sound undergoing some heavy audio processing to make it appear more mechanical. I suppose even if the best text-to-speech system had been used, the robot's jerky intonation would likely have detracted from the performance.[46] If the robot is meant to be artificially intelligent, the mistakes in the speech would break that illusion.

The audience seemed willing to overlook shortcomings in Baxter's appearance, movements and voice. He was essentially a high-tech puppet, with its motion being controlled by a robotics researcher just off-stage. My favourite part was Baxter waving around a skull and reciting 'Alas, poor Yorick!' But I was amused rather than moved by the robot Hamlet facing up to his mortality. Still, it whetted my appetite for more, so I sought out a full theatrical performance involving a leading robot.

*

RoboThespian and Judy Norman in *Spillikin: A Love Story*

In the play *Spillikin: A Love Story*, one of the main roles is played by the android RoboThespian. It sits in a wheelchair for the entire performance, with its mechanics exposed and a face created by projection. The voice is provided by actors reading the script because the play is set in the future where robots are capable of speaking more naturally. Of course the audience don't know this, and so the play allows us to glimpse a future where speech synthesis is completely natural. Alongside the voice, the theatre company has created a library of 200 preprogrammed moves that allows the robot to gesture when triggered by the script.[47] It certainly works. In the Q&A after the show I saw, a couple of audience members talked about finding humanity in the robot. As the creator of RoboThespian Will Jackson explained to me, 'A really good actor doesn't let you know that they're scripted, and neither does a good robot.' When I spoke to Will before seeing the play, he explained that he had created the robot actor to investigate people's willingness to suspend disbelief. This happens all the time in the movies and Will wanted to explore that phenomenon beyond a flat screen.

I asked him whether people found robots disturbing. 'Yes, of course they do, and that's part of the fun!' Will replied. 'The only failure is to be dull.' He believes that the audience has a powerful yet confusing experience because deep down they know a robot is purely mechanical and yet it is behaving with characteristics of a living thing. They start reading intentions into what the robot is doing, whether that is from the way it moves, how it looks or what it is saying. In short, they anthropomorphise it.

The main character in the play is a widow, Sally, who gradually descends into dementia. Played by Judy Norman, her decline is captivating, utterly convincing and harrowing to witness. Suffering from a terminal medical condition her husband Raymond created RoboThespian as a companion for Sally. The android is there to chat, reassure and provide memories when Raymond has gone. *Spillikin* raises many questions about the role of social robotics in society and whether technology should be used as a substitute for human companionship. Conversely, the robot also sheds light on what it is to be human. As carers we are imperfect and do not have infinite patience. RoboThespian is sometimes a poor carer because it displays human flaws that arise from the husband's programming. Unnervingly, however, there are occasions when a well-programmed robot will be a better, more patient companion than a human.

I asked the playwright Jon Welch about the possibilities of writing a play where robots were no longer playing themselves but were actually taking the roles of humans in a piece of standard repertoire. He began to describe in great detail all the elements that would have to be right, such as the voice, facial expressions, the pacing, etc. 'Obviously to a certain extent it is sacrilegious,' Jon said, 'but if you took that amount of trouble, I don't see why it wouldn't be entertaining and even moving.' We know that human actors are pretending but when they do it well we suspend disbelief and get engrossed in the story. Jon believes that this could also be achieved by robots, with the audience humanising them, starting to care about them and getting involved in the story.

But this is using robots as extremely complex and highly programmed puppets mimicking what could be created by human actors much more easily. And it would lack spontaneity. As Jon explained, 'Actors would hate it because part of the joy are the moments where you don't predict what is going to happen, and you get a bit of magic on stage that comes out of a mistake. Mistakes are often fruitful.' Also, why would the audience turn up to see a play 'robotically' delivered that lacks the excitement and unpredictability of a live performance?

The challenge to our sense of humanity would be even starker if the robots became more autonomous and behaved less like puppets. To achieve this, they would have to emote successfully using a text-to-speech synthesiser. Current artificial voices do not sound human enough, with the bottleneck being the process of interpreting and marking up the script with how the words should be delivered. If scientists managed to solve the problem of saying the lines convincingly, then AI would be at the point of being able to comprehend the script. At that stage even the playwright is likely to be redundant, because AI could in fact write the whole play. But despite the headlines about AI taking over the world we are a long way from achieving this. Just think about the totality of the life experiences that a human writer and actor can draw upon. As Jon Welch succinctly put it, a playwright has 'a universe in their head, and until we can put a universe in a robot's head, I don't know if that would be worth looking at'. Yet it hasn't stopped people trying, and I will return to the topic in the final chapter. But first, let's explore whether a computer can be a good listener.

7

Beware, The Computers Have Ears

How long will it be before we stop bossing our computers about and start having meaningful two-way conversations? For that to happen computers have to become better listeners. They have to go beyond an algorithmic decoding of speech and become attuned to our tone of voice. We have all experienced that sinking feeling when a loved one complains not about *what* was said, but about *how* it was said. For better or worse, vocal cues can reveal so much, whether that is our excitement when telling a funny anecdote, our boredom with a conversation, or our horror while recounting a disaster. A good conversationalist needs to pick up on these vocal cues, whether they are human or silicon. Can a computer be a good listener?

Of all computer-listening technologies, the most controversial ones are those for detecting deception. Most of us are deeply uncomfortable with the idea of being caught out by the analytic power of a cold, hard machine. Yet the prospect that technology could dis-

Lie detector test at the Clinton Engineer Works, 1944.

tinguish between truth and lies is extremely appealing to police officers and politicians wanting to protect the public from murderers, sex offenders, benefit fraudsters and other criminals. The lie detector has become a star on tabloid TV shows and is being hailed as the definitive arbiter of someone's sexual fidelity. And all this despite high-profile cases that have revealed the technology's fallibility.

The Green River Killer was named after the stream south of Seattle where he dumped his victims in the 1980s and 90s. One of the tools the police used as they hunted for the serial murderer was the polygraph. This is a machine that examines someone's truthfulness by focusing on physiological signs such as their heart rate, sweating and respiration. In 1984 Gary Ridgway, a family man who worked in a paint shop, volunteered to take a polygraph test and passed. Nineteen years later, Ridgway was jailed for forty-eight counts of aggravated first-degree murder when he was linked to the victims via DNA evidence.[1] The polygraph had evidently failed to identify the Green River Killer.

A scientific review of the polygraph by the British Psychological Society found that in criminal cases a test is correct in about 83 to 89 per cent of cases where the people tested are guilty, but only 53 to 78 per cent of the time when dealing with someone who is innocent.[2] Despite this, in 2014 the UK government started compulsory testing of high-risk sex offenders using the technology. Trials showed that being hooked up to a polygraph made sex offenders more likely to admit to risky behaviour like viewing pornography and meeting children. But in fact these disclosures were not picked up by the polygraph itself: the offenders confessed because they believed that the technology could detect lies.

If a polygraph is unreliable maybe we could train a computer to analyse speech instead? Voice-stress analysis is a controversial technique used by insurance firms, police forces and government departments to screen people for telltale signs of lying. ABC news claimed that the technique was used in Guantanamo Bay and Iraq before being banned by the Pentagon.[3] The companies which sell such systems do not disclose much about how they work but scientific studies have cast doubt on the efficacy of the technology. In contrast, there are standard ways of getting computers to listen to voices and interpreting what they hear that are well documented. These mainstream approaches have been applied to diverse situations including cars that detect whether its driver is drunk from slurred speech and a mobile-phone app that gives feedback to people with bipolar disorder about their mood swings.

Getting a computer to listen and make sense of what it hears usually means exploiting machine learning. This is where a computer program is trained to look at an audio recording and deduce useful information from it. Some important calculations in speech science are based on simple mathematical formulations. If you want to work out the frequency at which the vocal folds are opening and closing then there are equations to extract this information from the sound waveform. But if you want to deduce something more nebulous from an audio recording, such as whether someone sounds anxious, mathematical reasoning is unlikely to provide results. In such cases, a computer program has to 'learn' for itself through experience to spot the telltale signs.

The usefulness of machine learning for audio extends beyond speech. It has many uses in analysing music, for example in detecting the genre, whether a piece is classical, jazz, rock, etc. I worked with BBC R&D to investigate what emotions are being portrayed by the theme tunes of the broadcaster's TV and radio programmes. The BBC has a million recordings in its archives and the corporation wanted to tag each of these with its general mood – for example whether a drama felt light-hearted, sad or adrenaline-charged – so people could navigate the archive looking for items with a particular feel. Could analysing the theme music help? When the upbeat opening chords of the theme for the US sitcom *Friends* start playing, even if you have never seen the show before you can guess that this is a feel-good comedy. By contrast, most news bulletins start with a fanfare to set a serious mood. We wondered if a computer could learn to spot whether a piece of theme music was happy or sad, funny or serious.

Humans learn how to associate certain musical features with particular moods. Happy tunes tend to be faster and in Western music they often use a major key signature. Sad music is usually in a minor key with musical phrases that flow downwards to mimic the descending intonation that we adopt when delivering bad news.[4] We pick up these associations through a lifetime of exposure to music. Similarly, a machine-learning algorithm needs to build up an understanding by 'listening' to a large library of audio examples. One machine-learning technique that is currently undergoing a renaissance is artificial neural networks, the workings of which crudely mimic structures within the brain.

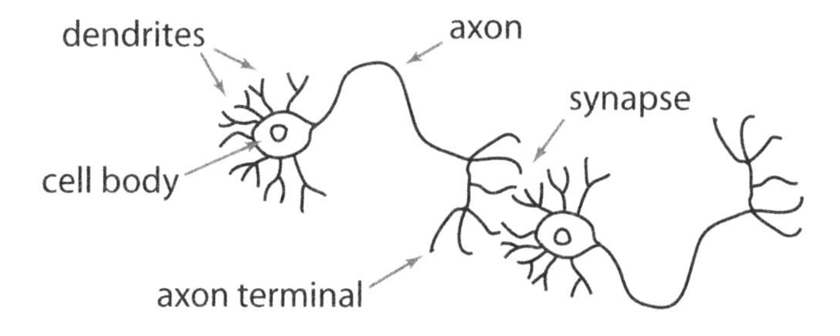

Two neurons.

The human brain is the ultimate learning machine. A baby's brain is made up of about 100 billion neurons, with each neuron connected to around 10,000 others. Each individual neuron has a relatively simple task. Information flows through it in the form of electrical impulses received by short-branched extensions of the cell called dendrites. The impulses are combined by adding or subtracting, depending on whether each connection is excitatory or inhibitory. If the combined signal exceeds a certain threshold then the neuron fires and sends another electrical pulse that races along the nerve fibre or axon. This then gets passed on to other neurons. What makes the brain so extraordinary powerful is the way that these simple neurons work together in vast and complex networks.

A child learns a new skill through a process of training. When a dad sits down and reads a book to his daughter the child's brain tries to associate the sounds she hears with the words that appear on the page. When the child starts to read the storybook for herself, the dad provides feedback on how well she is doing, congratulating her when she gets a word right and gently correcting her when it's wrong. This learning causes the strength, speed and number of the connections between neurons in the girl's brain to change. The child learns from the successes and mistakes, so that next time she reads a book she is more likely to get it right.

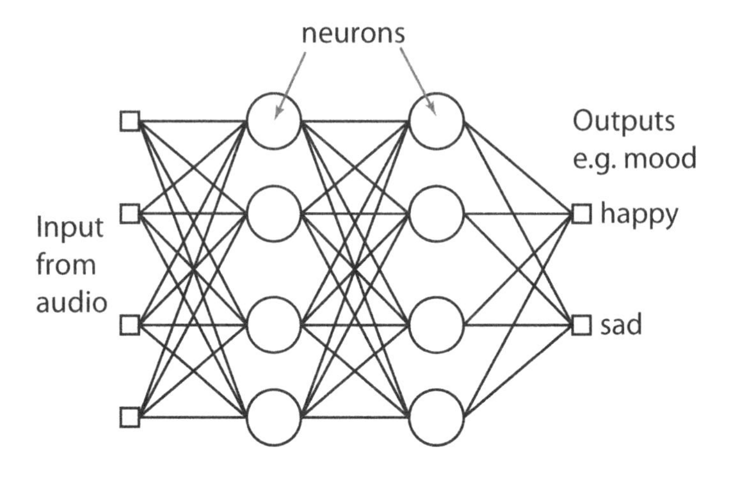

An artificial neural network.

Artificial neural networks try to mimic this type of behaviour. They are also constructed from a large number of 'neurons' that are able to do simple mathematical operations. Each artificial neuron is just a few lines of computer code which, like its biological equivalent, sums up and processes the input before sending on the result to other neurons in the network. These are not exact replicas of the neurons in the brain, however, and the number of connections is much smaller.

Like a child, artificial neural networks need training. A computer scientist acts as a surrogate parent, supplying examples to the network and giving feedback on whether the algorithm got the answer right or wrong. In order to train it to work out a theme tune's mood you can feed it recordings clearly labelled for whether the music makes the average listener feel happy or sad. As you might imagine, manually labelling lots of example tracks with the right answers is a tedious job. For the theme-tune work we turned to the public for help and ran an online experiment where 15,000 people listened to 144 theme tunes from the past sixty years and told us what mood each piece of music created. During training, the computer uses the feedback on how good its guesses of mood are to alter the strength of connections between the neurons. This is how the computer gradually improves its estimations. With enough examples, it gradually learns to pinpoint the emotion portrayed by a music track with more accuracy.[5]

That said, because artificial neural networks are nowhere near as powerful as the human brain, feeding them raw audio can overwhelm them. Humans have billions of neurons whereas large artificial networks only have thousands. The capacity for computer learning is therefore limited and so it's often better to simplify a task. In our case one way of doing this was to feed it a few carefully chosen features extracted from the sound rather than the raw audio.[6] Knowing that happy music tends to be faster, you could apply mathematical formulations to calculate the tempo of the music and input this to the artificial neural network. Another feature might be to determine which chords are prominent in a piece, which would reveal whether the music is composed in a major or a minor key and is thus more likely to be happy or sad.

What makes the machine-learning algorithm powerful is that once trained, it can make intelligent guesses about theme tunes it has never heard before. The system is far from infallible, however, and can only be as good as the data it learns from. When we expanded the work to try and guess the genre of the TV programmes from the theme tunes, some presented problems. One outlier was the dissonant chords that formed the plaintive tune to the children's programme *Noggin the Nog*.[7] As the music does not follow the upbeat stereotype for the genre, it stumped the machine-learning algorithm. But then it probably would have fooled a human as well!

Thus successful machine learning usually depends on identifying essential features that are revealing. To apply machine learning to detect lies, we need to know which aspects of speech would be useful to an artificial neural network trying to recognise deception. So what have psychologists discovered from experiments with humans? Are there telltale signs of a liar?

*

In January 1998, President Bill Clinton made his infamous statement, 'I did not have sexual relations with that woman, Miss Lewinsky.' His delivery was stilted and metronomic, emphasised by his fingers' rhythmic tapping on the lectern. Seven months later, the president went on national television to announce that he had lied. The contrast in Clinton's delivery of the two speeches is stark. His admission of lying is warm and fluent, delivered in that familiar style that helped make him a hugely successful politician. The starkest difference between the speeches is the change of rhythm: the second speech was no longer metronomic, with the pauses between the words having a natural variation.

When a parent is quizzing a teenager over where she was last night, a government agent is interrogating a suspect, or you are listening to a politician whom you suspect of lying, the assumption is that there will be some telltale signs that reveal the untruth. Being deceitful is usually stressful and we assume that anxiety or fear leaves a mark in the speech.[8] Stress heightens arousal and this makes it difficult to maintain precise control over the voice. In some people this affects the loudness, vocal harshness and the frequency at which the vocal folds vibrate. In others, the liar overcompensates for the effects of heightened arousal and delivery becomes overly precise. This would be a plausible explanation for Clinton's laboured delivery of his first speech.

The problem is that most of us believe that we are better lie detectors than we are liars. In fact, the truth is the other way round: we are much better at deceiving than we are at spotting it. This might arise from early memories of a high-stake lie that failed. The white lies where stakes are not so high – 'Sorry, but your email must have ended up in my spam folder' – will usually be less prominent in our minds. We forget we are actually quite good at deception. We tend to look out for certain signs that seem to indicate that a person is lying such as the talker looking away, smiling or fidgeting. But scientific studies show that these are poor indicators. Indeed, the idea that people are more restless when lying is the exact opposite of how liars tend to behave.

Eitan Elaad of Israel's National Police explored this in a deception study involving sixty police interrogators watching videos of teenagers and trying to spot when the youngsters were lying.[9] Eight teenagers were videoed describing people they liked or disliked. Sometimes they told the truth and sometimes they lied about their feelings. Getting such videos right is a real problem in deception research. The teenagers should appear to be stressed while lying otherwise there would be no cues for the police interrogators to work from but obviously there can be no genuine threat of jail or other severe sanction if a lie is uncovered. Still, there are various tricks to up the ante. You might appeal to the teenagers' self-esteem and tell them that only individuals who are highly intelligent, strong-willed and have excellent self-control can succeed at deception.[10]

Two-thirds of the Israeli police interrogators thought they did really well at spotting when the teenagers lied. In reality the policemen did worse than chance, with lies being spotted only 46 per cent of the time. They might as well have flipped a coin. On average, studies involving judges, psychiatrists and polygraph examiners find a success rate that is only a little better than guessing.[11]

In 1994, Richard Wiseman, professor of the public understanding of psychology at the University of Hertfordshire, conducted a large experiment into lying. It was less well controlled than laboratory studies, but Wiseman tested an unusually large number of lie spotters. It involved two interviews with the well-known British political commentator, Sir Robin Day. In one interview he lied, in the other he told the truth, and the public had to spot which was which. Over 40,000 people either heard Day on radio, read transcripts of the interviews in a newspaper or watched him on TV. Radio listeners who had verbal and vocal cues to detect the lies got it right 73 per cent of the time. The newspaper readers who were just given the text were correct 64 per cent of the time. Surprisingly, the television viewers who could both see and hear Day fared worst of all by doing not much better than chance at 52 per cent. It seems that the addition of visual cues actually decreased the ability to detect lies.[12]

In a worldwide survey, gaze aversion was the most common sign that people believed betrayed a liar.[13] Looking away is something children as young as five or six years old associate with lying. Why we rely on this wrong cue is curious because humans are generally very good at spotting somebody else's feelings. There seems to be an illusionary correlation. We avert our eyes when we are ashamed and being caught lying is a shaming experience. This may lead us to incorrectly believe that someone telling the truth will hold our gaze, while liars will look away.

The low success rate in deception studies is partly down to the reliance on stereotypical but erroneous cues such as gaze aversion. Another factor is that we have a truth bias: we naturally believe more statements to be true than false. In trying to detect a lie, as in much of life, we use heuristics – or rules of thumb – to make judgements and these are often subject to biases. Michael Shermer gives the following example in his book *The Believing Brain*.[14] Imagine you are an ancient ancestor who is out on the savannah and you hear a sound. Was that the wind rustling the undergrowth? Or was that a predator trying to creep up on you? Assuming it is an animal preparing to attack taking evasive action is the safest thing to do in the circumstances. It is dangerous to presume it is the wind when it might be a predator. It is to your advantage that you learn a heuristic with an inbuilt bias that every rustling sound on the savannah signals a predator.

In lie detection, the truth bias means we believe more statements to be true than false (except in some special circumstances, such as when we are listening to a sales pitch). A probable reason for the bias is that we come across many more correct statements than deceptive ones in our everyday lives. It is also often easier to corroborate what people say when they are being truthful. Lying is much less common, and when it does happen it is much more difficult to detect. Sometimes you might even unconsciously conspire to keep the deception secret. Does someone really want an honest answer when they ask if they look fat in their favourite dress? (This phenomenon has been dubbed the ostrich effect.) In short, we generally believe that we can detect both truth and lies although most of the feedback comes from truthful cases. We tend not to discover most of the lies we are told.[15]

There are some cues that can increase your chances of detecting lies but these can be subtle and difficult to spot. Liars tend to make more speech errors, appear sluggish when answering questions and speak more slowly. Extra thinking time is needed to plan and execute a lie. A truthful answer to a question is usually easier to produce because it reflects what we believe happened. There are many more options for camouflaging a lie, and if interrogated your brain works harder to keep subsequent answers consistent with previous lies.[16] Aldert Vrij from the University of Portsmouth believes that this can be exploited to make interrogations more successful. You might get a suspect to tell their story in reverse order, adding cognitive demands and so making it more likely that cues to lying are revealed.

One of the most famous liars of recent times is the former professional racing cyclist Lance Armstrong. He won the Tour de France an unprecedented seven times, but both during his career and after he retired he was repeatedly accused of doping. Standing on the Champs-Élysées with the Arc de Triomphe in the background, in his last victory speech in July 2005 he declared: 'For people who don't believe in cycling, the cynics and the sceptics: I'm sorry for you, I'm sorry you can't dream big, I'm sorry you don't believe in miracles ... there are no secrets – this is a sporting event and hard work wins it, so *vive le Tour* forever.' In 2013, Armstrong was stripped of his titles when he finally admitted to taking banned performance-enhancing drugs. What is striking about the old TV clips where Armstrong denies doping is how fluently and confidently he answers the questions put to him. This confirms research which shows that if you need to lie then try to keep your speech fluent to avoid being caught out.[17] Armstrong is a good example of another finding from deception studies: that some people who are very good at lying speak normally while doing so.

So are there any other vocal cues? Experiments have shown that the pitch of a liar's speech often goes up, with the frequency increasing by 6–7 Hz. There are several possible reasons for this, such as the stress of lying changing the heart rate, which then alters the sub-glottal pressure, which in turn quickens the oscillation of the vocal folds. Unfortunately a pitch increase is not universal. So in short, scientific studies have found no universal 'tell'.

Given this evidence, I find it surprising that we still tend to believe we are good at detecting liars. And doesn't it also contradict our personal experience of getting away with lying? (Not that I myself would ever tell a falsehood of course!) A survey of 1,000 American adults found that on average people lie 1.65 times a day, although many of the untruths were told by a small number of prolific liars.[18] Given that even the most trustworthy individuals tell fibs to avoid social embarrassment, why do we fail to learn from our own deceptions to get better at spotting when others are lying? The inconsistent and subtle nature of the vocal and verbal cues to deception makes this impossible.

Research in 2014 by Leanne ten Brinke and colleagues from the University of California at Berkeley suggests that while our conscious attempts to spot liars are hindered by looking at unhelpful stereotypical cues, our unconscious mind does slightly better.[19] Again, the researchers got subjects to watch videos of people lying and telling the truth. In this case it was a mock-crime interview about a theft of $100. In addition to asking whether the people in the videos were lying, ten Brinke used an Implicit Association Test in order to reveal the unconscious thoughts of the subjects.[20] Participants were asked to associate words such as 'untruthful' and 'honest' with pictures of the liars and truth-tellers, and the speed at which the word associations were delivered was then measured. Ten Brinke found that when the photos were incongruent with the word, for example when a picture of an actual truth-teller was shown alongside the word 'untruthful', the subject took more time to answer. So although the subjects were poor at spotting liars when simply asked to say who was lying, at an unconscious level they were picking up on cues as to the truthfulness of the speakers. If biases in our conscious mind get in the way of lie detection, maybe a dispassionate computer can do better.

*

Several commercial systems claim to detect deceit through voice stress analysis. In 2003 BBC News reported: 'A car insurer which introduced phone lie detectors says a quarter of all vehicle theft claims have been withdrawn since the initiative began.'[21] A year later, the *New York Times* reported the claims of one manufacturer that their technology was used by '1,400 law enforcement agencies across the United States, as well as by other state and federal agencies including the Defense Department'.[22] More recently, the Department of Work and Pensions in the UK spent £2.4 million on an evaluation of the technology involving nearly 3,000 benefit claimants.[23]

Many of these systems claim to work by monitoring 'micro-tremors'. Stress changes blood flow to muscles including those controlling the larynx, and the suggestion is that this alters micro-tremors in the voice.[24] Yet while studies have found small tremors in large muscles such as the biceps, there is no evidence that they occur in the laryngeal muscles. Controlling the voice is an incredibly complex process that uses the smallest and fastest muscles in the body to allow rapid articulation. Even if micro-tremors were present, their effects would be undetectable.

The scientific failings of voice stress analysis were detailed in an academic paper published in the *International Journal of Speech, Language and the Law* by two linguistic and phonetic experts from Sweden, Francisco Lacerda from Stockholm University and Anders Eriksson from Gothenburg University.[25] The paper did not disguise the authors' contempt for the technology. 'Charlatans may be found in all walks of life,' the introduction began, 'especially in activities where there is a possibility of making money, and forensic speech science is no exception.' The paper was controversially withdrawn from the journal's website after one of the companies whose technology was ridiculed threatened to sue the publisher. The case was one of the reasons that the UK libel laws were changed in 2013 to protect scientists who publish peer-reviewed material in academic journals.[26]

The paper focused on one particular patent for the technology, which confirmed the authors' fears about the technique. 'It read like a student essay,' Francisco told me. 'A student who didn't understand what it was about, and is just [using] some fancy wording.' The patent disclosed a 500-line computer program enabling

Francisco to reconstruct the detection process. The program takes the wiggles of a sound waveform from a voice recording, processes them and then works out the number of peaks, troughs and plateaus. Plateaus might be caused by pauses, 'ums' and 'errs', and so might have some weak correlation to the fluency of speech. But the number of peaks and troughs in a waveform is hugely influenced by the settings of the audio recorder.

'It's like taking a text, counting the number of incidences where you have a vowel between two consonants, and then you measure the number,' Francisco explained, 'and the length of the sequences of characters that are within say five or ten steps in the alphabet. On the basis of this, you tell me what the state of mind of the author [is]!' Francisco described the program as a 'voice controlled, quasi-random [number] generator'. Based on the number of peaks, troughs and plateaus, it outputs a series of labels such as 'Untruthfulness, Low stress, Thinking less than in the calibration, Normal excitement'. As Lacerda's and Eriksson's paper describes, the 'output of an analysis is structured much along the same lines as horoscopes', producing a pattern that the operator can interpret almost at will.

Such systems have also been subjected to scientific tests that show them to be no better than chance. In one study, Kelly Damphousse and her colleagues from the University of Oklahoma asked 319 detainees in Oklahoma County Jail about their use of drugs and put their answers through voice stress analysis.[27] After the interview, urine samples were taken and analysed to reveal the truth. The study showed that only 15 per cent of liars had been detected by the voice stress analysis. More worryingly, for every real liar the technique falsely identified nine truthful respondents as not telling the truth. 'False positives' in machine learning are really important. Imagine taking up one company's suggestion of using voice stress analysis to screen passengers at Heathrow airport: each day you would have 8,000 irate innocent travellers falsely identified as security threats.

In another prison study, the number of detainees who lied dropped by two-thirds when they were told that their speech was being analysed.[28] Thus voice stress analysis only seems to work because of a bluff: people are less likely to lie when they think they will be found out. Psychologists call this the 'bogus pipeline effect'. The phenomenon was discovered by Edward Jones and Harold Sigall who used a fake lie detector to make test subjects 'open up a pipeline to their soul' and so discover people's true beliefs.[29] Police, insurance firms and the government could save lots of money by pretending to buy the technology! Still, it makes me wonder how long such a bluff could last.

It only takes a quick search on the Internet to find evidence that these systems do not work. But voice stress analysis is a zombie technology. However much it gets knocked down by scientific evidence, it somehow rises again. Ignoring the research, the UK's Department of Work and Pensions spent £2.4 million between May 2007 and July 2008 trialling the use of the technique to cut benefit fraud. The idea was that when claimants rang a government agency, the voice stress analysis would alert staff to callers who needed more focused attention. In four out of seven trials, covering 80 per cent of the phone calls examined, the system did no better than flipping a coin.[30] 'It's a pity they took such a huge amount of money to come to that conclusion, when they could have just asked the relevant questions to begin with,' Francisco Lacerda told me.

More generally, the problem with assuming that stress is a sign of lying is that both the liar and truth-teller might be in a stressful state. Deception researchers call this 'the Othello error'.[31] In Shakespeare's play, Othello accuses his wife Desdemona of having an affair with Cassio, his lieutenant. Cassio had previously been seen with a handkerchief that Othello had given to Desdemona. Othello overhears what he assumes to be the successful assassination of his lieutenant, and tells Desdemona that Cassio is dead. She thinks she has no chance of proving her innocence. Her anguish is taken as further evidence of guilt by Othello and he kills her.

If Othello was alive today, could a computer help him determine Desdemona's guilt or innocence? As someone who has worked in machine learning for many years my bet is that examining Desdemona's intonation and speech rhythm alone is unlikely to tell him the truth. If all the scientific studies into deception have failed to find any clear patterns in the way people lie, and if the voice can be altered by stress even in those telling the truth, then even the best machine-learning algorithm is going to struggle.

What about a seemingly more simple task: could computer listening estimate how intoxicated a talker is? When we are drunk our speech can alter dramatically. Speaking requires an immensely complex coordination of fine motor movements. With enough alcohol imbibed, muscle control can lead to clumsy and slurred speech as we struggle to manipulate our vocal anatomy. We might also speak slower because of problems with articulation and sluggish cognition.

Forensic voice analysis was at the centre of the court case against Joseph Hazelwood, the captain of the *Exxon Valdez* oil tanker who was accused of being intoxicated when in charge of his vessel. The oil tanker ran aground off the Alaskan coast in 1989 and the accident led to the escape of 41.8 million litres of crude oil, killing roughly 250,000 birds, 3,000 sea otters, 300 harbour seals, 250 bald eagles and twenty-two killer whales.[32] Recordings of Hazelwood at the time of the accident showed that his voice had changed. He spoke more slowly and there were variations in the roughness of his voice.

Could a computer detect such changes in a captain's voice and automatically pass on the command of a ship to the first mate? In 2011 researchers took part in a competition to explore how well computers could detect intoxication from voice recordings.[33] The first step was to prepare a selection of examples for the researchers to work on. This was done by getting 154 people tipsy and asking them to recite phrases. Research groups were then challenged to develop a computer algorithm that could determine whether each audio example indicated that the talker was intoxicated or sober. The best program achieved an accuracy of 71 per cent.[34] This is similar to the success rate that you or I might achieve: on average humans can spot intoxicated speech three-quarters of the time.[35] Unfortunately, the success rate for the computer is too low to be a reliable way of vetting captains.

In the *Exxon Valdez* court case, although Hazelwood admitted drinking vodka before boarding the vessel, he was acquitted. One reason was that the voice analysis could not unambiguously prove he was drunk. The changes to his speech could have been caused by him raising his voice to be heard on the noisy ship.[36] While computer listening can pick up information from the voice, like humans, the judgement is fallible either because the algorithm is imperfect or because the vocal cues are ambiguous.

*

Lie-detecting algorithms have so far overlooked the words in speech. Maybe a computer would be better at detecting intoxication if it looked out for specific phrases such as 'You're my best mate you know' or noticed how drunk people the word wrong order often get? Jonathan Aitken was a high-achieving UK politician, once tipped as a future Conservative prime minister. In 1995, when he was chief secretary to the Treasury, Aitken resigned from the Cabinet to fight allegations made by the *Guardian* newspaper and Granada TV. They claimed he had accepted bribes from Saudi businessmen connected to weapons sales. His speech announcing his libel action did not hold back: 'If it falls to me to start a fight to cut out the cancer of bent and twisted journalism in our country with the simple sword of truth and the trusty shield of fair play, so be it. I am ready for the fight.' Four years later, Aitken was jailed for eighteen months for committing perjury and perverting the course of justice. During the libel action he claimed that his part of a £1,000 bill from the Paris Ritz Hotel was paid for by his wife using money he had given her. The *Guardian* got hold of a copy of the bill, however, and the lie was revealed. Aitken's political career was over. Listening to archive footage of his 'sword of truth' speech, his delivery is unnaturally flat in comparison with the vitriolic words.

For a computer to have a chance of spotting this lie, therefore, it would have to be able to understand the words. This would allow a detection system to use other signs of deception that have been revealed in scientific studies, such as the fact that when we tell lies we give fewer details and make fewer links to external events.[37] But to exploit this, a computer would need to be able to recognise speech and understand the semantics.

One of the first electronic speech recognition systems, called Audrey, was built in 1952 by K. H. Davis and colleagues at Bell Telephone Laboratories in the US. It could recognise individual digits and with careful adjustment for a particular speaker it could identify virtually all the words correctly. Like many early systems, Audrey was essentially based on matching patterns. The picture on the next page shows traces of someone counting slowly from one to five. The top shows a common way of representing sound, a wiggly trace indicating how the pressure created by the voice varies as the five digits are said. The second utterance 'two' shows two distinct phrases, 't' and 'oo'. It starts with the plosive consonant 't', where air is first stopped by resting the tongue on the roof of the mouth, and when the tongue then drops a fast puff of air creates the sound. This is quickly followed by the vowel 'oo' that is almost sung. The lower plot is a spectrogram that shows how the frequency content of the speech varies. For the word 'two', a dark line slopes downwards from left to right, whereas for 'three' there is a diagonal dark line going in the opposite direction. As the speaker says the second part of 'three', his intonation creates an increase in frequency, and hence the rising line in the spectrogram.

The spectrograms are like fingerprints and reveal that each digit has a unique pattern. Audrey's task was matching a pattern from the sound spoken into a microphone to one of the expected patterns for each digit. In the 1950s this task was challenging because of the lack of computers to create spectrograms. Moreover, Audrey was not a particularly practical system – as James Flanagan from Bell Labs noted, it 'occupied a six-foot high relay rack, was expensive, consumed substantial power and exhibited the myriad maintenance problems associated with complex vacuum-tube circuitry'.[38]

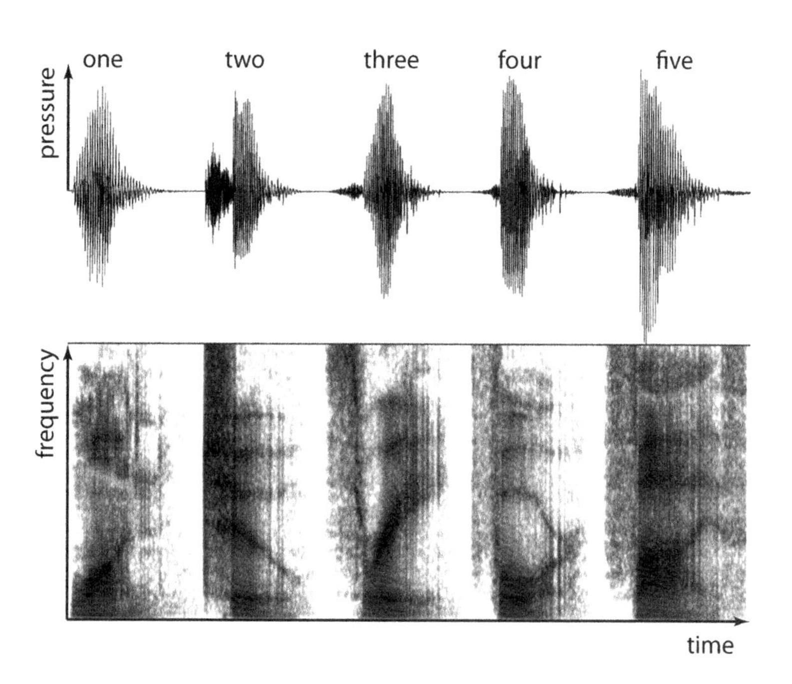

A male voice counting 'one, two, three, four, five'.

The other problem with this type of analysis is that we pronounce words inconsistently. For instance, a word that normally would be pronounced with a drooping frequency might get a rising intonation if it ends a question. And pronunciation varies even more between individuals, so your spectrogram for counting from one to five would look different from mine. Even the best modern systems using much more sophisticated techniques than Audrey get caught out. When the iPhone 4S launched in the UK in 2011, the voice-activated assistant Siri struggled to understand strong Scottish accents.[39]

In recent years, a neat confluence of faster computers and the use of machine learning have halved error rates for speech recognition. While still some way from being as reliable as a human listener, modern systems no longer require you to speak slowly with long pauses between words. What's more, in an era of big data, systems are trained on a vast number of examples. This is how Apple solved the problem with Siri: the computer listened to more Scottish so it could learn the accent. Big data also means that speech recognition systems have a vast vocabulary, with Google claiming about 3 million words, which is much bigger than any human's. This is why speech recognition works even if you talk about a very specialist subject with its own niche vocabulary.

Nowadays we are all generating vast amounts of digital data while shopping, using social media or browsing the Internet and giving companies huge amounts of information about ourselves in exchange for free services. Allowing computers to tap into our voices makes the data even more valuable because on top of the words there is potential to find out about how we are feeling.

However, applying machine learning to large corpuses of data can have unintended negative consequences. You might think that because they are built on mathematics and algorithms they would be as objective as Dr Spock from *Star Trek*. But the computer software learns to replicate social prejudices contained in the training data. In 2017, Aylin Caliskan and colleagues from Princeton University looked at associations between words in a popular database used to train machine-learning algorithms.[40] The database contained billions of words hoovered up from the Internet. In one test Caliskan examined which personal names appeared in sentences close to pleasant words such as 'love', and which names were near to unpleasant words like 'ugly'. The results showed racial prejudices, with European American names being more often close to pleasant terms than African American names. Another test showed gender biases with male names being more associated with words relating to work like 'professional' and 'salary', and females close to terms describing family life like 'parents' and 'wedding'. Train a machine-learning algorithm on such a database and you risk creating software that is sexist and racist.

Bias is already evident in popular tools like Google Translate. Use it to translate the Turkish phrases '*o bir doktor*' and '*o bir hemşire*' into English and you get 'he is a doctor' and 'she is a nurse'.[41] Yet '*o*' is a gender-neutral third-person pronoun in Turkish. The presumption that a doctor is male and a nurse is female reflects cultural prejudices and the skewed distribution of gender in the medical profession: we have ended up with a translation algorithm that is sexist. Using such an algorithm to screen medical job applications would reinforce existing cultural biases. Although discussions of AI often centre on the algorithms being used, it is often the data that determines the performance and can lead to unintended and disturbing consequences. In 2015, Flickr released an image-recognition system that incorrectly tagged black people as 'apes', and pictures of the concentration camps at Dachau and Auschwitz as 'jungle gym' and 'sport' respectively. Without care, similar mistakes are going to happen when computers try to identify human characteristics from our speech. Especially as our voice contains subtle but often ambiguous information about our race, sexuality and gender.

Companies such as Google, Apple and Microsoft now have vast databases of sound recordings that they employ in creating speech recognition systems. In one experiment, Microsoft used twenty-four hours of data from their voice search app that contained over 30,000 utterances. People were looking for businesses, so typical words included 'Walmart', 'McDonald's' or '7-Eleven'. After training, the artificial neural network achieved a sentence accuracy of 70 per cent on voice enquiries it had not previously heard.[42] Such a success rate is impressive considering that these search recordings would have come from people with different accents and included mispronunciations and background noise. It still means, however, that many of the words being suggested by the algorithm are wrong. But this is not a problem unique to computers. As we have seen, when humans hear speech there are often bits that are missing or wrong and our brain fills in the gaps or makes corrections. The same applies to reading. It's not that difficult to make out the following sentence: 'Aoccdrnig to a rscheearch ... it deosn't mttaer in waht oredr the ltteers in a wrod are, the olny iprmoetnt tihng is taht the frist and lsat ltteer be at the rghit pclae.'[43] Corrupted text can be recovered provided enough of the letters are correct, and the same is true of speech.

When you start to type a query into a browser, several suggestions for the rest of the search text appear. When I type 'Trevor Cox' into one search engine, the first suggestion is 'Trevor Cox whl', as I share a name with a Canadian ice hockey player who plays for the Medicine Hat Tigers. These suggestions are possible because big data is being used to create models of language, in this case words that are likely to go together in search terms. Such language modelling is vital for speech recognition as it allows wrongly identified words to be corrected.[44]

Voice search is surprisingly effective but can it help detect lying? Not at the moment, because the language model focuses on likely search terms, about which Google has vast amounts of information. The company has started to analyse the number of false facts on a webpage so rankings of research results can be based on the trustworthiness of a site.[45] But this can only go so far in helping with lie detection because written and spoken language work differently. Just consider the richness of word play in something like a spoonerism and the problem of developing a language model to work with it. Born in 1844, clergyman William Spooner had trouble with his tongue and brain keeping up with each other. At a wedding he reportedly said, 'It is kisstomary to cuss the bride.' And he once accidentally suggested a toast to 'our queer old dean' instead of to 'our dear old queen'.[46]

Scientists have tried to use machine learning to detect jokes including double entendres.[47] They get the computer to look for words that have lewd overtones like 'banana'. There is also a particular structure to erotic sentences that some double entendres share, such as '[subject] could eat [object] all day'. After training, the computer could detect double entendres about 70 per cent of the time. (This sentence is crying out for a double entendre about hard problems in machine learning.)

Computers might have a better chance listening out for the distinctive sound of laughter to spot a joke. When I met the UCL neuroscientist Sophie Scott to ask about impressionists we also discussed her research into how we express emotions. Her work started with testing screams and people going 'yuck', only later switching to the more pleasant task of investigating laughter. She has had to convince sceptics that this is a serious subject for study. An anonymous colleague of hers once pinned this note to some consent forms that had come out of a communal printer:

> This pile of paper appears to be rubbish, based on the contents* and will be disposed of if not collected.
>
> *is this science?

But laughter is a serious subject because it is a default state for us. 'All other things being equal, you're feeling comfortable and OK with the people you're with. You'll laugh around those people,' explained Scott. If laughter is missing it is a sign that something is wrong. An extreme example comes from people with gelotophobia who fear laughter because they believe it is ridicule directed at themselves. 'It's 100 per cent correlated with someone being in a pretty full-on psychotic state,' was how Sophie described the condition. Researching laughter helps get to the heart of social interactions because it lubricates conversations. Couples who abate the inevitable stresses of being together through laughter are more satisfied with their relationships and stay together longer.

Before discussing the unique sonic signature of laughter, Sophie reached for a model of a brain so she could point out the different parts involved in listening. For speech, the left hemisphere tends to be involved in the phonetic, semantic, lexical or syntax information. This means the right hemisphere focuses on everything else in a voice like intonation or identifying the talker. Therefore, when Sophie puts someone in a scanner and plays them laughter, the right side of the brain shows more activity.

Before Sophie embarked on the brain scanning she needed a good set of laughter recordings. Her first recording sessions with a couple of her researchers went fine; in fact they had 'a phenomenally good time making each other laugh'. But when they tried to harvest some laughs from a second group of volunteers it fell flat. 'It didn't occur to me until the first poor man [from the group] was sitting on his own in the anechoic chamber, and not laughing, that of course no one knew each other [well] and they were not a group of friends,' Sophie told me. The laughter donor needs to be on their own in the anechoic chamber, so that Sophie can get a clean recording of their voice. Yet laughter is a social activity. So Sophie and her researchers had to come up with a new protocol. It started outside the chamber where they would 'spend a lot of time sitting together, watching videos, everyone laughing together, warming people up', Sophie explained. 'And then finally you get someone up and running, and you throw them in the anechoic chamber.'

When Sophie played these recordings back to volunteers in a brain scanner, she found that there were two types of laughter with distinct neurological responses. Laughter is a natural response to something funny, but the most common type is the polite social laughter that oils conversations and mostly has little to do with humour. These posed laughs are signalling engagement and enjoyment as you chat to someone, and in a ten-minute conversation there are typically five of them.[48] When people hear posed laughter, Sophie noticed more activity in the medial prefrontal areas of the brain that are used to work out people's intentions. For a social signal like laughter, it seems logical that theory-of-mind networks should get involved in decoding the sound.

The other sort of laughter is when you really let go and laugh uncontrollably. 'Ha ha ha' is a very simple vocalisation. Each 'ha' is created by spasms of the diaphragm and intercostal muscles forcing puffs of air out of the lungs that then set the vocal folds in motion. For real laughs there is higher pressure, which in turn creates a higher pitch, than we find in posed laughter. You also get other fry and whistling sound resulting from the uncontrolled use of the vocal anatomy.[49] Uncontrollable laughter can be a very odd sound. The comedian Jimmy Carr is an extreme example and has described his laughter as sounding like a happy dolphin – unusually he actually laughs while inhaling.[50] When listening to such an uncontrolled laugh, the brain shows strong responses in the left and right auditory cortex, which are located just above the ears. As a real laugh sounds very different from speech, singing or other everyday sounds, its novelty leads to more activity in the auditory cortex.[51]

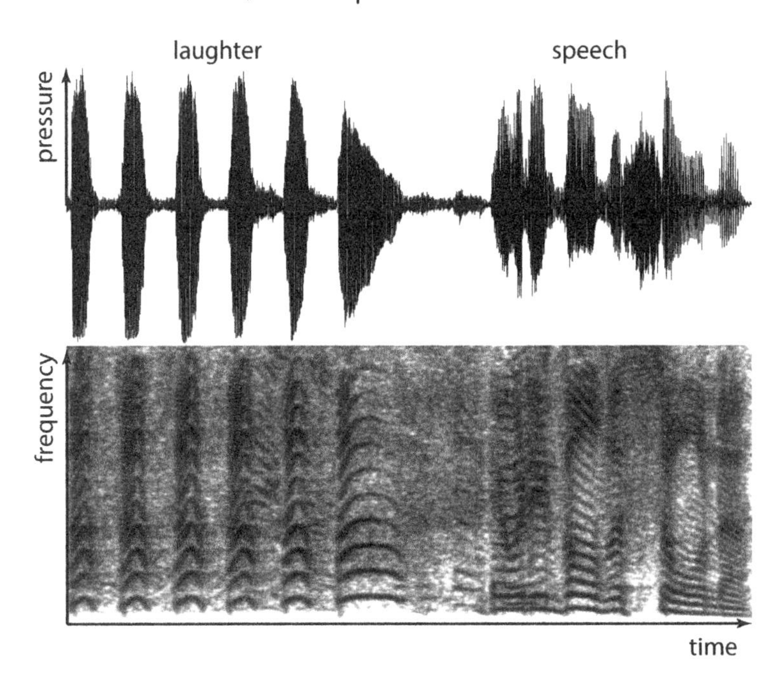

Jimmy Carr laughing ('ha ha ha ha ha ha ha') and then talking ('that would involve a sauna').

As laughter has a very distinct acoustic signature, computers using machine learning can pick it out with a very high success rate.[52] Unfortunately this does not mean that a computer is able to detect humorous speech. As most laughter does not occur in response to an actual joke but serves as a social lubricant, the computer is often going to get it wrong. To be a good listener to detect jokes and lies, we need a computer to know much more about language. Currently, computers have learnt a vast amount of simple information by rote, but they have no idea of what is really going on.

*

The fictional civil servant Sir Humphrey Appleby in the two BBC series *Yes, Minister* and *Yes, Prime Minister* once said, 'A good speech isn't one where we can prove that we're telling the truth – it's one where nobody else can prove we're lying.'[53] One trick to make a lie hard to notice is to embed it in an otherwise truthful statement. A criminal might make subtle alterations to a true story, such as changing the time at which an event occurred, allowing him to speak mostly the truth while giving a misleading alibi. Omission is another tactic. Faced with an outfit including a garish shirt, a partner might comment on the cut of the suit jacket and omit commenting on the hideous pattern underneath.

Confronted as we are with a multitude of deception methods, no wonder we struggle to be a reliable lie detector. People have developed many different approaches to lying because deception is a crucial skill that comes with evolutionary benefits, as is evident in primates who hide food and covertly mate. We have all embellished a story to make it more fun and memorable. And white lies are an important part of social group interaction.

For humans, learning deception is a marker in child development. About 25 per cent of two-year-olds have the ability to lie; this rises to about 90 per cent in four-year-olds; and by the age of eight virtually all children can lie.[54] This is an important indicator of brain development. Children who start lying earlier are showing faster cognitive development – something that parents faced with a lying toddler might find comforting. For in order to lie, a child needs to conceptualise how others perceive information.

Computer simulations of how populations interact show that in societies based on cooperation and honesty, an individual can gain an advantage by occasionally being deceptive and lying, provided that the gains are high and the risks of being discovered are low.[55] The simulations also show it is important that deceivers get found out some of the time to ensure that cooperation dominates; hence a fallible ability to detect lies is not a shortcoming, it is an important part of how societies develop.

Studies have shown that primate species that display more cooperation also show more deception behaviours. Humans dominate the world by collaborating with each other. Vocal signs of lying are subtle, complex and inconsistent because there is an evolutionary advantage if we deceive occasionally and get away with it. Evolutionary pressure would make a species of Pinocchios learn how to control their noses. Given that humans cannot infallibly detect liars, and we have more sophisticated listening skills than a computer can currently muster, it seems unsurprising that voice stress analysis falls short. To get a reliable lie detector requires deep artificial intelligence to analyse the speech and voice even better than humans can do.

8

Computer Love Letters

Computers are dumb servants dutifully running code as directed by their human programmers. Novelists use word processors, sound engineers mix music on computers and cartoonists use animation programs, but the machine is not being artistic: creativity is a uniquely human ability. Yet scientists are challenging this notion by coding computers to write poetry, compose music and devise new cooking recipes. So can machines be creative? Opinions are divided, with some believing that creativity emerges from biological consciousness and therefore electronics will always fall short of what humans can do. Others argue that consciousness and creativity are ultimately mathematical processes, which in theory could be done by a machine even if in practice they are too complicated to codify. Whatever your view on this topic, there are already computers aping narrow aspects of human creativity by penning news reports, proposing and testing scientific hypotheses and painting artworks. Meanwhile, by simulating creative processes in computers, researchers are revealing what lies behind human inventiveness.

Artificial intelligence is going to have a profound effect on speech. We have seen how computers are learning to listen and talk. But what about that vital ingredient of communication: the words? Automatic transcription tools are already breaking down language barriers between humans and allowing conversations that previously would have been difficult to conduct. Half a billion people use Google Translate each month and, perhaps unsurprisingly, requests to translate 'I love you' and 'You have beautiful eyes' are most common.[1] But computers can do more than facilitate online chat by translating sweet nothings.* A true artificial intelligence will not just parrot memorised phrases, it will be creative and take language in new directions.

We have a dark fascination with the idea that machines might use their own creativity to exponentially grow their abilities, quickly outstripping humans and taking over the world. This explains the sensationalist headlines in 2017 when Facebook's AI laboratory closed down two chatbots because they started deviating from English as they talked to each other. While some called the development 'incredibly scary' because the machines could 'bypass its masters', wiser heads pointed out that the computers had simply found a more efficient way to communicate. This is what happens with language – it evolves. The chatbots were meant to be learning how to negotiate better with people but the researchers forgot to restrict them to just work in English.

To understand what computers might be able to do, let's reimagine Edmond Rostand's nineteenth-century play *Cyrano de Bergerac*. The original has the brilliant poet Cyrano sporting a huge bulbous nose, a disfigurement that prevents him professing his love to the

---

* For example the creation of chatbot therapists that mimic talking therapies. There is evidence that some people are more willing to open up to a piece of technology than to a human therapist.

beautiful Roxane. The handsome but dim Christian is Roxane's other suitor. The two men come to an arrangement whereby Cyrano writes poems for Roxane but Christian claims to be the author. The tragic romance ends with Roxane discovering that Cyrano is the true poet just moments before he dies. In my version of the play, I envisage Christian bypassing Cyrano, and turning to a computer to write the poems. The code in the computerised 'Cyrano' would be driven not by love, but by an algorithm that seeks to maximise Roxane's poetic pleasure.

Here is an example of a computer love letter:

> DARLING SWEETHEART
> YOU ARE MY AVID FELLOW FEELING. MY AFFECTION CURIOUSLY CLINGS TO YOUR PASSIONATE WISH. MY LIKING YEARNS FOR YOUR HEART. YOU ARE MY WISTFUL SYMPATHY: MY TENDER LIKING.
> YOURS BEAUTIFULLY
> M. U. C.

On that evidence the original Cyrano would have little to worry about, but these sentences are of historic importance: they are early examples of computer-generated text. They were written by 'Baby', the world's first modern computer with memory to store a program; 'M.U.C.' stands for Manchester University Computer. In 1952, printouts of strange love letters appeared on a university noticeboard. They were the brain child of computer pioneer Professor Christopher Strachey. He had gained a reputation as a formidable coder following his audacious running of a vast punchcard program that played draughts. It astounded everyone by executing almost faultlessly first time; it finished by playing 'God Save the King'.[2] Strachey wrote the love-letter program to test Baby's ability to randomly select information. The program rolls a dice to choose romantic verbs and nouns, and places them into a template to make the sentences. It lives on today as part of an art installation, *Love-Letters 1.0.*, created by the artist and media archaeologist David Link, which runs the original source code.[3]

While the program was an impressive achievement for its time, Strachey and not Baby was the creative source behind these sentences. Moreover, no one would mistake the printouts for a love letter written by a human. Here is a recent attempt by a computer to complete one of Shakespeare's sonnets, starting from the famous words 'Shall I compare thee to':

> Shall I compare thee to skill,
> The white heart of the release moon still come and say,
> And all the one rain is the our heads, in the green day.
> The precision of countrys of the corner so little garden,
> darkening blessed of sea while one of the city for the voice in the window –
> The rusty ban, who deadest an old and the starpy called people,
> And the draw of the doors and the time for proval, so many instant
> Than are a river on the warm of the blood of his willing and play
> And gave me here some move; and what any.*[4]

Again, Roxane is likely to be underwhelmed if Christian recited this semi-literate stream of consciousness. In this case, however, the computer programmer is much more detached from the creative process. In 2015, James Lloyd and Alex Davies, then of Cambridge

* Not all the words in this poem are real and there are grammatical and spelling mistakes. The reason for this will become clear when the generation method is described.

University, trained an artificial neural network on 10,000 poems. Like the algorithms that convert speech to text which we encountered earlier, these programs learn to do a task through being exposed to numerous examples. The neural networks were set up to predict how word patterns will evolve over time. So when given 'Shall I compare thee to', it makes a guess at how the line of poetry might end, and then continues to predict how the sonnet would continue to evolve letter by letter.

During training, the output of the neural network is compared to real poems. If the network incorrectly predicts the next letter, it updates its internal connections to improve future forecasts. The system does not have the capacity to rote learn every poem exactly, so the predictions are never the same as the original verse.

Working letter by letter means the algorithm has to learn even the most basic structures of a language. After a few seconds of training, the program creates gibberish and yet it has learnt that 'e' occurs frequently in English: '/Wteh lea e a sti es s e inne re l se l lhre, so e sir a f e riay r mn rdh rewsr e iie r eto e ctsse e i o en e tnea e s'.

After a few minutes of training, the neural network has worked out that letters clump together into 'words':

> ursoe haoth sicge tim bonr
> ghoiconiiroch is a)o
> PuTTY dhr doooc nins voaed ofitot tions anewt

And after five minutes, some of the words were even proper English:

> Stand the fanes and chen the posser.
> Srone the she was insoneed the crour faning of mas,

Over the next few hours the non-words became less common and the grammar improved:

> Are you not pleasant?
> And as I am leaving you my life like the earthworms?

The final sonnet is not great poetry, but it is astonishing that such a simple piece of machine learning can produce anything that is recognisably verse. If you give a computer free rein you risk the poetry being disappointing; constrain what the program can do and it is more likely to produce a verse you might enjoy. Take a limerick, for example. With its well-defined form it is much easier to get a computer to produce something reasonable. It is even possible to create poems which readers will struggle to identify as being written by a machine. If you wanted to explore this you might conduct a bastardised Turing Test, for which you select a number of computer and human poems and challenge readers to guess the authorship. Here are some lines of poetry, one created by a computer and the other written by a human. Can you tell which is which?

Poem 1:

> By action or by suffering, and whose hour
> Was drained to its last sand in weal or woe,
> So that the trunk survived both fruit & flower.

Poem 2:

> nuclear Parisian age
> as last as a proclamation
> last like a proclamation!
> as close as an interest!

The English Romantic poet Percy Shelley wrote the first verse. The second was produced by a computer program to which we will return again later. A simplified Turing Test is fun to do, but it is a poor test of creativity. I can easily influence the outcome by picking terrible examples of human poetry and so making it harder to spot the computer verse.

Alan Turing was the father of modern computing whose brilliance helped decode messages from the German Enigma machines in the Second World War. He also worked on Baby, the Manchester computer that wrote the love letters. The test he invented and which is named after him is often held up in the media as a vital threshold for artificial intelligence, much to the annoyance of many computer scientists. Turing wanted to find out whether a computer can think like a human. In a seminal paper he wrote about an 'Imitation Game' where in addition to composing poetry the computer would also need to critique the verse. As an example Turing gave the following hypothetical conversation:[5]

Interrogator: In the first line of your sonnet which reads 'Shall I compare thee to a summer's day', would not 'a spring day' do as well or better?

Computer: It wouldn't scan.

Interrogator: How about 'a winter's day'? That would scan all right.

Computer: Yes, but nobody wants to be compared to a winter's day.

Interrogator: Would you say Mr Pickwick reminded you of Christmas?

Computer: In a way.

Interrogator: Yet Christmas is a winter's day, and I do not think Mr Pickwick would mind the comparison.

Computer: I don't think you're serious. By a winter's day one means a typical winter's day, rather than a special one like Christmas.

Current computer poetry programs would be unable to deal with this interrogation because answering interview questions is a different skill from writing verse. At this point in time creativity displayed by machines always occurs in a narrow domain: the ability to work more widely is presently beyond artificial intelligence. But maybe we should not be too critical, as many human artists also struggle to explain their creative intent and processes.

Turing's paper also discusses 'contrary views', however, and quotes Sir Geoffrey Jefferson, then the chair of neurosurgery at the University of Manchester. In response to Baby, in 1949 Jefferson wrote about the dangers of 'anthropomorphising the machine':

> Not until a machine can write a sonnet or compose a concerto because of thoughts and emotions felt, and not by the chance fall of symbols, could we agree that machine equals brain – that is, not only write it but know that it had written it. No mechanism could feel (and not merely artificially signal, an easy contrivance) pleasure at its successes, grief when its valves fuse, be warmed by flattery, be made miserable by its mistakes, be charmed by sex, be angry or depressed when it cannot get what it wants.[6]

As art is the expression and communication of human experience, Jefferson argued that it could not be created by a computer – although as we have seen in the context of theatre, machines can shed light on humanness.

Many researchers focus on the process of making a piece of art, looking for insight into creativity and sidestepping having to deal with the subjective opinion of fickle human critics. The poetry-generation system developed by Joanna Misztal-Radecka and Bipin Indurkhya of Jagiellonian University in Krakow is one example.[7] Their verse is created by the computer mimicking some of the complicated processes of the brain.

Imagine getting together with colleagues at an awayday and being tasked to collaborate in writing a poem on a flip chart. The verse must be inspired by a blog post. Each colleague focuses on a specific task. One might identify the inspiring keywords in the blog's text, looking for words that have lots of associations and therefore are fruitful for making poetry. Another would evaluate the blog's mood, whether it conveys anger, fear or happiness. Someone else uses a thesaurus, writes possible synonyms and antonyms of the keywords onto some Post-it notes, and sticks them on the flip chart. Yet another person's task is to find words that trigger emotions which reflect the mood of the original blog. This pool of ideas is then taken by others who generate snippets of verse. They might create metaphors, oxymorons or rhetorical questions and then scribble them on yet more Post-its. Someone else is asked to collate them and select the best; they might check, for example, whether the lines scan. Eventually a poem will emerge.

Joanna's algorithms mimic this process with each person's task being performed by a piece of software. Here is an example of the output. The poem was inspired by the following blog:

> I remember being endlessly entertained by the adventures of my toys! Some days they died repeated, violent deaths, other days they traveled to space or discussed my swim lessons and how I absolutely should be allowed in the deep end of the pool, especially since I was such a talented doggy-paddler.[8]

The topic selected by the software was 'deep end' and the emotion 'anger'. So here is the poem:

I knew the undisrupted end
I was like the various end
As deep as a transformation
O end the left extremity
Objective undisrupted end
I hated the choleric end
O end the dead extremity

When Joanna and Bipin got people to comment on the quality of poems such as this, most verses did not rate very highly. This is hardly surprising because the computer lacks a way of properly evaluating the worth of the lines. In fact, if you found anything profound in the poem then this would be a happy accident. Yet poetry is an art form where much of the meaning is in the mind of the reader. So how was your reading of the poem influenced by knowing that it was computer-generated? There are studies from music that can shed some light on this.

In a scientific paper published in 2008, Nikolaus Steinbeis and Stefan Koelsch from the Max-Planck Institute for Human Cognitive and Brain Research in Leipzig described a study that looked at people's brain responses to music using an fMRI scanner.[9] The participants auditioned music by the twentieth-century composers Arnold Schönberg and Anton Webern. Some of the time the researchers pretended the composer was a computer, at others listeners were told it was written by a human being. Atonal music was used because listeners might naively accept the seemingly random notes as the output of a computer. When they believed the music to be written by a human, then those brain areas playing a role in predicting what others might be thinking showed increased activity. These results, alongside a questionnaire taken after the brain scans, showed that listeners were trying to figure out the intentions of the composer when it was believed to be a person. Atonal classical music is a niche interest; it would be fascinating to extend the study to more mainstream music, especially electronica which is full of synth sounds and electronic effects.

Presumably the same thing would happen at a poetry slam with robots performing machine verse. You would not attempt to work out the intention of the author for works produced by a computer. Literature is often autobiographical and draws on a writer's own lived experience. How can a machine hope to know what it is to be alive? Computer creation cannot be ascribed to a struggling artist writing about his troubled life. However romantic that notion may be, it cannot currently be coded into a computer program. We believe that any computer verse is going 'to lack soul' and it is therefore unlikely to recruit theory-of-mind networks in our brains.

One way to mitigate this is to get computers to repurpose or build upon text that is rich in humanity because it is authored by real people. This is what many of the entrants to National Novel Generation Month do. Created by Internet artist Darius Kazemi, it was inspired by National Novel Writing Month, an event where hundreds of thousands of writers try to create the first draft of a 50,000-word novel in just thirty days. Although started in the US, the event now attracts participants from around the world. This process of hurried human creativity has led to bestsellers such as Sara Gruen's *Water for Elephants*. The computer event challenges programmers to write code that automatically generates a novel of at least the same length. In 2015, there were a couple of hundred entries.

When I met with Darius the following year, he told me that some of the best moments in computer-generated works occur when the code produces text that stumbles as it 'tries' to be human.

The computer might sound like a tourist mangling a foreign language, and it can be fun to see how it fails, generating accidental epiphanies in the process. Darius's favourite entry came from 2014 and is called *The Seeker* by Thricedotted.[10] It is in fact one of the few entries that might be readable from start to finish. It uses a clever conceit: the story is about an artificial intelligence learning how to be human by reading Wikihow articles. Hence, as the protagonist is a machine, the reader does not expect its English to be perfect, and the bulk of the text can be sourced from a corpus of human writing. Much of the book consists of statements that are more like computer code than conventional sentences. So the novel starts by looking at the Wikihow entry on 'Getting girl to ask you out' and lists advice extracted from the text like, '01 … ALWAYS (PRACTICE_GOOD_HYGIENE) => good.' Every fourth page contains a short piece of generated prose that reads like surrealist verse. Such a book might appeal to fans of experimental literature, but you would have to be very dedicated to read the whole text. Even Darius admitted he has not read every page.

While computers might struggle to produce a whole book, they have already created thousands of short reports for Associated Press. Here is a snippet from AP's coverage of a Minor League Baseball game:

> STATE COLLEGE, Pa. (AP) – Dylan Tice was hit by a pitch with the bases loaded with one out in the 11th inning, giving the State College Spikes a 9–8 victory over the Brooklyn Cyclones on Wednesday.[11]

While the text is impressive because it reads like normal English, this straight relaying of facts is hardly going to take over from great sports writing. There is much talk of AI taking jobs, but journalists having nothing to fear in the short term.

Computers currently struggle with the subtle plotting and narrative required to sustain long prose. 'I don't think I've successfully generated 5,000 words that are compelling all the way through without extreme intervention on my own part,' Darius admitted. Take a classic whodunnit. A human writer will sow suspicion and insert clues about the murder throughout the book as well as laying false leads and red herrings. It is not the overarching narrative that makes a good book. The basic plot of *Murder on the Orient Express* by Agatha Christie can be summarised in a few hundred words, but it is the complexity of the narrative – the portrayal of the characters, the twists and turns, dead ends and gradual revelations – requiring tens of thousands of words that made the book such a bestseller. Computers are not going to replace Agatha Christie, explained Darius. But he believes they might compete with William Burroughs.

When I saw the android Bina48 being interviewed while exploring robot theatre, the disjointed conversation sounded like it came from a piece of experimental literature. How could Bina48 be made to hold a more natural conversation? Some researchers think that it would require computers to better understand storytelling. Mark Riedl and colleagues from the Georgia Institute of Technology crowdsource ideas for how a narrative might develop, and then use this information to construct flow diagrams for how a scene might unfold. Riedl believes that computers need narrative intelligence, the ability to tell and understand stories, if they are supposed to relate to humans.[12] Everyone at some point has shouted at their computer in desperation because it seems to have developed a malevolent attitude; this happens because the machine has no comprehension of what the user is trying to achieve. If we are ever to converse fluently with a machine then it requires narrative intelligence. And a computer could then learn how to behave by reading books about etiquette, social norms and values. After all, we implicitly teach and learn behaviour through stories, from the cautionary tales told to children at bedtime, through to romantic novels exploring the ups and downs of relationships.

But even if a computer read all the books in the world its knowledge would be incomplete. One problem is that the 'message' of the story is often implicit. Take a classic morality tale. It is much more powerful to leave the reader to draw their own final conclusion because the learning will be stronger due to the mental effort required to work out the underlying message. Or, to take another example, some of the pleasure derived from movies like *Donnie Darko* is working out what went on by chatting to friends as you leave the cinema. Some stories are deliberately told in a convoluted and ambiguous way to make them more intriguing. Then there's the detail that storytellers omit and leave to our imagination. A novelist describing bank robbers fleeing the scene of their crime leaves out most of the facts. There might be a clichéd wrestle with the ignition key in the getaway car to add tension, but many of the other details like opening the door, sitting down in the driver's seat, closing the door, etc., are omitted because they are unnecessary and boring.

Last but not least, great stories usually focus on the unusual. Without their own lived experience, computers who learn about human society through literature will have a warped view of real life. You may imagine that a conversation with a future computer might be like talking to the worst pub bore who recounts a story about the drive to work that was completely uneventful. In fact, the opposite is more likely, the computer will tell tales that just are too fantastical.

Getting the right balance between the mundane and the extraordinary lies at the heart of creativity, whether that is a dad making up a bedtime story, a sports commentator narrating a game or a comedian performing improv. Explorations into whether computers can achieve this are most advanced in music. As an abstract art form, it is easier to program a computer to create a pastiche of a musical style than write prose. As a teenager, I devised a computer program that composed ragtime music using simple probability tables. For example, if the current note is an A, then what is the probability that the next note is B, C, D, etc.? The notes were selected by rolling a dice. I then superimposed the structure and rhythms of ragtime on these melodies. This produced music with the jaunty lilt of ragtime but without any direction or resolution. It certainly was not going to match the quality of any of Scott Joplin's compositions – but then that was also true of my attempts to compose using pen and paper. Current music-composition algorithms use more sophisticated techniques than those I hacked together in my bedroom. The best have even created music that has been performed by professional orchestras in concerts halls. The computer program 'Emily Howell' composes in the style of Mozart or Beethoven and you can buy CDs of the music, although I think it is unlikely that the works will still be performed a hundred years hence.

It is easy to give a computer a snippet of melody and tell it to write some variations, say in a baroque style. Student composers go through this sort of exercise to develop their skill, but we would not think they were being especially creative. And even if there are programs producing music that I would find difficult to distinguish from real pieces by Johann Sebastian Bach, why do we need a machine to do this when Bach has already done it? This is impressive but mere mimicry. The computer program is not going to go against current musical trends and create something completely new and enthralling – there will be no punk-rock rebellion.

One approach to machine composition is evolutionary computing, where a program produces music by imitating the process of natural selection. In nature, chromosomes carry the genetic codes that shape life, with the genes evolving over time in response to evolutionary pressures. In evolutionary composition, the musical score can be thought of as the music's chromosomes, with each individual note being a gene. In nature, evolution needs a diverse range of individuals so that over generations the genes that aid survival are gradually selected. In an analogous way there needs to be a population of many different melodies to create a diversity of musical genes. As the computer software runs, new generations of musical scores are born and old ones die off. The best musical examples are most likely to pass on their genes to the next generation. As happens in nature, genes of each offspring come from their mother and father, so a new musical score is an amalgamation of the parents' tunes. For example, the start of the phrase might come from one parent, and the end from the other. Amalgamating melodies means that the population loses diversity, however. To counter this, mutations are part of the breeding process. Every time an offspring tune is born, there is a small chance that a mutation will randomly change a note in the score to increase genetic diversity.

In real-life natural selection, the parents most likely to breed are the ones best adapted to the environment. For evolutionary composition there must be an evaluation of the musical value of the tunes to allow the best parents to be selected for breeding. In GenJam, a piece of software that plays jazz, the quality of each computer improvisation was determined by the system's programmer Al Biles, drawing on his knowledge as a trumpeter. When Anna Jordanous from the University of Kent evaluated three different music-composition systems for her doctorate, GenJam scored highest. It came across more like a live performer and was less formulaic. This 'gave it a slight air of humanness', Anna told me. If you look online you can find a performance by Biles on trumpet and the computer on saxophone synthesiser.[13] It includes a call-and-response improvisation of the jazz standard 'Lady Bird' and there is a great moment on the video where Biles smiles at the computer's improvised response to his own playing. The computer does a reasonable job at being a middle-of-the-road improviser. Anna hopes that computers might advance human creativity. She wants music software to be 'seen as something that musicians can learn from, or be inspired by. Or even be very critical about, and learn by criticising what [the computer] does badly.'

Watching Biles and his computer duetting, the interesting question is how creative is the machine? The same question will arise when computers get better at writing prose. But it is difficult to answer. Philosophers have struggled to find an answer to the question 'what is art?' and creativity is no easier to pin down. The criteria would be different for a child's 'what I did during the summer holiday' story compared to a literary novel, for example. Yet, both child and novelist are displaying creativity. Anna thinks that we 'judge computers a bit more harshly because we find the idea of a computer being creative uncomfortable'. When being judgemental about computer art, we overlook the poor works that even the best human artist occasionally produces (my favourite example is the incredibly cheesy 'Delilah' by Queen).

According to Maggie Boden, professor of cognitive science at the University of Sussex, 'Creativity is the ability to come up with ideas or artefacts that are new, surprising, and valuable.'[14] A monkey bashing a typewriter will produce a random stream of letters that will be new. But it is unlikely that many words will emerge, let alone a work by Shakespeare, even if you had many monkeys and a lot of patience.[15] It is easy to program a computer to simulate the monkey experiment, but no one is going to read the stream of random letters that would appear in the hope that something worthwhile would magically emerge. Creative ideas need evaluation and one solution would be to get humans to laboriously score the merit of every work created by a computer, say on a scale from one to ten.

Boden suggests three ways of generating a surprising idea, revealing three types of creativity. The first is unfamiliar or unlikely juxtaposition. Take Salvador Dali's *Lobster Telephone* from the 1930s. In this famous surrealist object, the artist covered the handset with a lobster. Many viewers appreciate the similarities in shape between the two, along with the absurdity of putting them together.[16] But there was a deeper meaning, as for Dali both lobsters and telephones had sexual connotations, and the crustacean's genitalia were now next to the mouthpiece. The second type of creativity is exploratory, and this is what jazz musicians do when they improvise: they explore a conceptual space. Although constrained by the genre they can nevertheless generate new musical motifs, sounds and rhythms. This is what GenJam does. The third is the deepest type of creativity that stretches conceptual spaces to create astonishing ideas that previously seemed impossible. James Joyce wrote *Finnegans Wake* with words formed from snippets of different languages. Published in 1939, it still divides opinions today: is it a masterpiece or unreadable nonsense? Whatever your assessment, it broke new ground and freed future writers from having to follow strict rules of narrative, vocabulary and structure.

This third form is the hardest to achieve, and consequently attempts to mimic creativity in computers generally exploit the first two types. This was true of a brave experiment in computation creativity that aimed to produce an entire piece of musical theatre using computers.[17] The show was commissioned by Sky Arts and filmed in 2016 for a two-part documentary, *Computer Says Show* – a title that could have come from a piece of software trained to write punning TV programme titles. The musical itself was called *Beyond the Fence*. The setting was the anti-nuclear protests at the Greenham Common peace camps in the 1980s, with a slow-burn plot centring on a romance between single mother Mary and US airman Jim. The *Guardian* gave the West End show a two-star review:

'This software-generated Greenham Common musical is risibly stereotypical but pleasant as a milky drink.'

It might not have reached the dizzy heights of Boden's deepest type of creativity, but Geraint Wiggins, professor of computational creativity at Queen Mary University of London, told me that 'it was quite a good musical'. He admitted to having a 'tear in my eye at one point', although he suspected that it was due to the human input from the writer, composer and performers rather than output from the computer. Still, Wiggins thought that the software which generated the premise for the show marked a huge leap forward.

The musical's basic plot was the product of the 'What-if Machine' created by Maria Teresa Llano and colleagues at Goldsmith University, which tries to model the creation of fictional ideas.[18] It takes possible themes, main characters and plots and combines these into pithy sentences that describe a scenario. For example, the theme might be aspirational, the character a soldier and the plot a quest. One suggested scenario would be 'What if a soldier needed to avoid a confrontation in order to clinch a victory?' A usable idea, although a bit of a cliché.

One approach to constructing these scenarios is to subvert normal associations. As Mark Riedl commented in *New Scientist*, 'Narrative psychologists often say that a story is only worth telling when there is a breach of convention.'[19] Take for example the fact that dogs like bones. The What-if Machine changes this to create the scenario 'What if there was a little dog who was afraid of bones?' Here we have an example of Boden's theory that creativity can arise from unexpected juxtapositions. Another approach is to use synonyms to exaggerate. The statement 'people enjoy jumping' becomes much more intriguing as the question 'What if people were addicted to jumping?' But while the What-if Machine impressively encapsulates some of the creative approaches that we use when making up a story, there are aspects that are missing. The machine is unable to work across modalities, for example. Mendelssohn's overture *The Hebrides* was inspired by the choppy seas and misty views on a ferry voyage in western Scotland.[20] How might a computer mimic this ability to work across domains? Humans achieve this because language allows us to bridge between different senses, our memories and our emotions. A computer algorithm needs an equivalent language to unlock its creative potential.

Another difficulty for the What-if Machine is that it lacks sensibility and struggles to produce something valuable. 'What if there were a little cat who couldn't find a litter box?' might make a trite children's story, but this scenario does not conjure up many fruitful narratives. There is a sweet spot for ideas, ones that are fresh enough to be intriguing, but not so random as to be unfathomable. For the musical *Beyond the Fence*, human intervention was needed with the composer and the writer sifting 600 scenarios and selecting the best. They chose 'What if a wounded soldier had to learn how to understand a child in order to find true love?'

The musical was a love story from the 1980s, featuring a female protagonist and a happy ending. This remit was created by another piece of software written by the machine learning group at Cambridge University. They did a statistical analysis of 17,000 musicals examining what ingredients featured in the most successful ones. Along with another piece of code, this produced a structure for the musical. It identified the need for an opening number to hook the audience – the equivalent of 'Willkommen' from *Cabaret* – and a catchy comedy number, like 'Officer Krupke' from *West Side Story*.[21] This sort of statistical analysis is also done by humans; 'I think that's what brains do,' Geraint told me. If someone regularly goes to musical theatre, they have knowledge of how a musical should be structured, although much of it might be unconsciously known. This creates expectations for the narrative arc, such as the need for the 'I am what I am' song when a character sings with defiance and triumphalism.

What made *Beyond the Fence* particularly groundbreaking was how it brought together different aspects of creativity. The lyrics were created by the same program that attempted to complete Shakespeare's 'Shall I compare thee'. The two scientists working on the code, James Lloyd and Alex Davies, gave the program lyrics from musicals to learn from, but the resulting lines were like a stream of consciousness from someone who keeps wandering off-topic.[22] Again a human lyricist intervened and selected the best material. It was the same with the melodies, where a musician picked out tunes created by the computer and set them to fit the lyrics and expand them into full songs.[23]

To produce musical songs properly, say an upbeat number, the computer has to do more than pick some happy words and a jolly melody in a major key signature. At best such an approach would end up with a simplistic children's song. Music is generally much more nuanced. The hits of the English singer-songwriter Lily Allen provide some good examples. Tracks where the mood of the music jars with the sentiments conveyed by the lyrics are one of her specialities. In 'Not Fair', Allen sings bitterly about a boyfriend's inability to perform in bed, with an accompaniment that would not be out of place in an upbeat Eurovision song. A computer needs to become a better listener to improve its songwriting: it needs to understand how the melodic arc of the music changes the speech prosody, how the lyrics will be perceived, and how this supports the narrative intent. We are a long way from having a computer that can detect subtle vocal cues for sarcasm and irony, let alone one that can compose a song with these features.

Although decades of work might be needed to solve these issues, the research effort is already producing tools that are being used by musicians today. The FlowComposer from Sony is an interactive composition tool where AI makes the first stab at creating a score before a human musician tweaks and adjusts it to create the final piece.[24] This collaboration is unlikely to replace great songwriters and composers, but there are plenty of routine composition jobs where this could be useful; it could more quickly produce a backing track to cheap corporate videos, for example. AI will also be useful in education. Algorithms can be trained to give feedback as humans learn new skills, whether that is a music student starting to improvise or a budding orator perfecting their charismatic voice.

*

Artificial intelligence and humans working together on creative endeavours is also going to become increasingly common outside the arts. Take software engineering, for example. Writing a computer program is an exercise in problem-solving: what instructions do I need to give to a machine to achieve a particular task? But complex programs are laborious to code and prone to human error. Like the FlowComposer that supports musicians, researchers are now developing tools to support human software engineers. At its most simple, programmers might benefit from a process that resembles predictive text, where new lines of computer code are automatically created without a human having to type it in. The algorithm works out what bits of code should come next by analysing a vast database of other computer programs.[25]

Other researchers are exploring how computers might code independently. Google has given their machine-learning algorithms working memory to get closer to what humans have available in the brain. Google's Neural Turing Machines can learn to produce procedures that perform simple computing tasks, but so far these have only learnt extremely basic ones, such as repeatedly copying and sorting data. But although only in its infancy, AI already allows computers to solve some defined problems better than humans. For example, Google Search now uses machine learning to improve performance, whereas in the past humans would handcraft all the rule-based algorithms that created the page rankings.[26] Computer self-coding is in its infancy, but it will be revolutionary. If you consider recent radical and disruptive technologies, like the iPhone, most have software at the heart of the product.

Within many other branches of engineering it is becoming increasingly common that final designs are actually determined by a computer rather than directly by a human. I have used such software myself to create treatments that improve the sound of theatres. As we have seen in Chapter 5, a well-designed acoustic helps an actor's voice to reach the back of the audience. If you get the material, shape and form of walls and other surfaces right, then the pattern of sound reflections will enhance rather than hinder speech. I specialise in designing lumpy surfaces that are called diffusers and disperse sound. When a diffuser is applied to a large flat wall it is a bit like applying a frosting to a mirror. If the visual image in a frosted mirror is blurred, similarly the acoustic image is less distinct when sound reflects off a diffuser. This can help with sonic aberrations such as echoes from the rear walls of theatres.[27]

When I started working on diffusers the best designs drew on clever mathematical principles. My innovation was to use a computer to search for surface topographies that produced the right acoustic reflections and had an appearance that fitted with modern architecture. This is done through a process of computer-based trial and error. We have already seen how new music can be generated by mimicking the rules of evolution. The same process can be applied to acoustic engineering.

Top row shows two 1970s diffusers. Below is a wavy ceiling in a Cinerama that I designed whose curves better fit current interior-design fashions.

Was my computer being creative? There is a test for artificial intelligence named after the nineteenth-century English mathematician Ada Lovelace. Lovelace is often called the first computer programmer, because of her detailed critique of Charles Babbage's analytical engine. This machine was the first ever design for a computer. It could calculate mathematical functions following a program, but as Lovelace commented, it 'has no pretensions whatever to originate anything. It can do whatever we know how to order it to perform. It can follow analysis; but it has no power of anticipating any analytical relations or truths.'[28] The Lovelace test is about an artificial intelligence creating something that defies the explanation of the human creator of the computer code. To achieve that, my computer would need to have hypothesised that mimicking the rules of evolution could lead to better diffuser designs, and then designed the scientific experiments necessary to prove that to be true. My software would fail the Lovelace test.

Science, engineering and maths need innovative ideas and artefacts to progress. An artist builds on the canon of previous artworks; similarly a scientist 'stands on the shoulders of giants', building on current knowledge and understanding. Ultimately, both artists and scientists have to come up with something new, surprising and valuable. Scientific enquiry is one of the pinnacles of human achievement, but even this is not immune from the advance of creative computers. This new approach to science is most advanced in biology. At the University of Manchester, Professor Ross King and others have created a robot scientist named Eve that they hope will help with the discovery of new drugs. King took me on a tour of his startlingly white laboratories, including the small room where Eve works. She looks like a small industrial robot with two arms to pick up and manipulate samples. All around her are stores of chemicals, incubators and cameras. These allow Eve to set up experiments, grow cell cultures, photograph the results and use image analysis to work out how well the cells grew.

Eve automates the tedious experimentation needed when developing new drugs and can examine 10,000 compounds a day. The hope is to find drugs for diseases such as malaria and African sleeping sickness. Laboratory automation is common because robots are better at pipetting samples and can work around the clock. But Eve is much cleverer than a brute force trial-and-error machine. She does not just try every imaginable combination of chemicals in the hope of stumbling across a useful drug but hypothesises scientific theories, and then designs and executes experiments to explore these ideas, including updating her knowledge based on what she finds.

To allow Eve to do this she needs knowledge of the field within which she works. The first robot scientist that King worked on was called Adam, who was equipped with a model of metabolism in yeast and basic textbook knowledge of chemistry. (Yeast is studied because it resembles human cells.) One advantage of using artificial intelligence to explore drugs, however, is that computers can have a more extensive and detailed knowledge of a particular area than a human scientist. Unfortunately, much of this knowledge exists in scientific papers and translating that to a form usable by a machine is difficult and time-consuming. But this problem is gradually being overcome. In one study, IBM's Watson, the computer that famously won the US game show *Jeopardy!*, analysed 70,000 scientific articles on a protein tumour suppressor called p53. Based on what it learnt from reading the scientific literature, Watson identified six new proteins for testing in the laboratory that could modify p53.[29]

How do Adam and Eve expand their knowledge and carry out science? They use deduction, abduction and induction. Ross explained deduction to me using a variant of a classic example from Aristotle. Given the two facts 'some birds are swans' and 'all swans are white', it is possible to deduce that 'some birds are white'. This sort of logical reasoning underlies many areas of computer science. Abduction and induction are more interesting because they are how the computer can generate a scientific hypothesis that is to be tested. Induction is where general rules are drawn up from a set of examples. If Aristotle were to observe birds in Greece then he might infer that all swans are white. But this induction is wrong, as a visit to Australia where there are black swans would demonstrate. Abduction is what Sherlock Holmes does, coming up with the most plausible explanation from a series of observations.* 'All swans are white', 'that bird is white' and therefore 'that bird is a swan' is an example of abduction. The hypothesis that the bird is a swan could then be tested by further observation. (Sherlock might find that the white bird is in fact a goose.) Similarly, once Adam constructs an abduction about yeast, the computer designs the best experiments to test the hypothesis, and then sets about doing the work using the robotic arms and other paraphernalia. Cultures are grown in different conditions and photographs used to quantify how well the cells grow. These results then reveal whether the abducted hypothesis is likely to be correct, and so allows the theories in Adam's memory to be updated. Using these processes, Adam discovered new scientific knowledge about which genes make certain enzymes in yeast.

Where does Adam rank in terms of inventiveness in science? 'It's not enormously creative, it's a very simple sort of science; it's subhuman in many ways in what it can do,' explained Ross. 'It's superhuman in other ways, because it knows all the literature, it can do pipetting much better than humans.' There is one disadvantage, though. 'What it can't do, for instance,' Ross told me, 'is redesign its representation of the problem like a human could.'

Computers play a central role in most scientific exploration but we are now entering an era where machines will become more than a dumb servant to ease the tedium of scientific endeavours. By combining the best of human creativity with machine-learning tools, science can be made to advance more rapidly. Ross believes it will go even further in the future with computers being able to do science better than a human. Unlike the arts, applying artificial intelligence to science is not hampered by the complication of dealing with human value judgements. 'Nature is honest ... the world isn't

* Holmes is often described as using deduction, but that is actually incorrect – he abducts.

trying to fool us,' Ross explained. 'It's an objective thing whether the computer is generating new science or not.'[30]

Artificial intelligence has potential to change speech by enabling new science that leads to new technologies. Like Edison's phonograph, these could revolutionise talking and listening.

*

In the debate on creativity, a distinction is often drawn between what is new to a particular person and what is new to the world. Only Einstein came up with the theory of relativity – a true historical first. But everyone is creative, has novel thoughts and finds new solutions to everyday problems. I have just worked out how to get more dirty crockery into my new dishwasher. Such creative thinking has no historical significance but the solution I invented was new to me. Naturally historians concentrate on the world-firsts and the groundbreaking artefacts produced. But creativity is a process and how it plays out in the everyday is important. It is not the preserve of a romanticised elite but a normal part of human intelligence. Being inventive about how to catch prey, preserve food to avoid starvation, or protect a settlement from attack might not be very artistic, but these creative abilities help explain why humans have come to dominate the world.

While it might be fun to see if a computer can write poems like Cyrano, the everyday stories that humans have told around the fire or dinner table are arguably more important than literary art. It is the quotidian activity that allows knowledge about how to survive and thrive to be passed between us. These conversations enable humans to transcend the slow process of biological evolution, allowing rapid cultural and technological evolution. As the author Philip Pullman once said when asked about why stories are so important to us: 'Because they entertain and they teach; they help us both enjoy life and endure it. After nourishment, shelter and companionship, stories are the thing we need most in the world.'[31]

While a computer is not going to produce anything as good as Cyrano de Bergerac anytime soon, most humans would also fail to reach such literary heights. However, getting computers to mimic the processes in the brain gives us insight into human creativity. As Geraint Wiggins explained to me, creativity draws on the need for the brain to continuously predict what is going to happen next. This has obvious survival advantages: our defence systems are anticipating what might wait around the corner, what dangers might be looming, and matching our alertness with what can be heard, seen and smelt. The brain therefore always has to strive to improve the quality of the predictions based on successes or failures in the past. We have evolved a strong response to our predictions being wrong and are constantly updating our forecasting.

Our memory cannot store an exact replica of what it senses because of the sheer amount of information. Even if we had the capacity to hold all the details, retrieval would be too slow. That is why a memory is actually a reconstruction based on some approximate representation of past events in the brain. As a result, your treasured first childhood memory might actually be a fabrication based on stories told in the family. Memory is dynamic. The brain is continuously working out the best compact representations of information for efficient storage and effective predictions.

No single forecast is perfect. What makes the brain powerful is that there are many predictions happening in parallel, playing out scenarios based on different assumptions. If we were consciously aware of all these then we would be overwhelmed. Consequently, most of this happens unconsciously. There is a process that selects the best prediction to attend to, and only then do we become consciously aware of it. Such a model explains the 'Aha' moments of creativity, where ideas appear to arrive from nowhere and are suddenly fully formed in the brain. In fact, they did not come from nowhere: they arrived from the unconscious mind predicting the future. It might also explain why when faced with a difficult decision, a good tactic is to do something else for a while, before making a final choice.[32] During the distracting task, your brain can unconsciously ponder solutions.

When Edison recorded and played back 'Mary Had a Little Lamb', listeners would have been continually predicting what would come next: the next phoneme, the next word, the next line of the verse.[33] A prediction of how a discourse might unfurl is important because it allows us to deal with mispronunciations or misheard parts of the speech; for a phonograph recording it allows the brain to estimate parts of the speech hidden by the crackles from the tinfoil cylinder. Geraint Wiggins demonstrated the importance of prediction to me by halting abruptly in different places while we chatted: 'Take a simple sentence and.' It is a subtle torture because your brain wants to know what happens next but cannot be sure because the rest of the sentence is impossible to predict.

Humans make predictions at many levels, ranging from what is the next phoneme all the way up to working out the implications of what is being said. To do this we need a large brain with a huge network of interconnected neurons. Neuroscience is only just beginning to glimpse which brain regions are involved in creativity and the complexity of interconnections.[34]

If creativity arises from predictive processes that afford an evolutionary advantage, could this be simulated in a computer? And could such a model go further and help explain how language evolved? Could it cast new light on the language ability of ancient hominins? Scientists are now configuring computers to mimic how the brain is organised: this will enable us to play what-if games to explore these questions. We might be able to explore how a proto-language could have evolved and better understand the role creativity played in the evolution of speech. Computer creativity has come a long way. It might have started with love letters to test the capabilities of the world's first modern computer, but in the next few decades it might reveal how our remarkable ability to talk evolved.

Acknowledgements

I would like to thank the large number of people who helped with the book, including: Daniel Aalto, Jamie Angus, Christella Antoni, Robert Asher, Matthew Aylett, Naheem Bashir, Peter Bell, Tam Blaxter, Fabian Brackhane, David Britain, The BSA delegates, Patrick Campbell, Jen Chesters, Christian Bech Christensen, Deborah Cox, Jenny Cox, Michael Cox, Nathan Cox, Peter Cox, Stephen Cox, Helena Daffern, Bill Davies, Nikki Dibben, Rachel Everard, Bruno Fazenda, Charles Fernyhough, Sue Fox, Gareth Fry, Jörg Hensgen, Jos Hirst, Nick Holmes, David Howard, Bipin Indurkhya, Anna Jordanous, Darius Kazemi, Simon King, Simon Kirby, Francisco Lacerda, Adrian Leemann, Mark Lewney, Louise LePage, David Link, James Lloyd, Sophie Meekings, Duncan Miller, David Milner, Joanna Misztal-Radecka, Judith Newman, John Potter, Louisa Pritchard, Steve Renals, Sophie Scott, David Shariatmadari, Dan Stowell, Johan Sundberg, Peter Tallack, Ingo Titze, Rami Tzabar, Aldert Vrij, Anna-Sophia Watts, Oliver Watts, Eloise Whitmore, Geraint Wiggins, Stuart Williams and Tim Wise. And apologies to anyone who I have accidentally missed off this list.

Lyric Credits

p.74 and p.75: 'the end is near' ... 'lived a life that's full' from 'My Way', lyrics by Paul Anka.

p.128: 'oh baby baby' from '... Baby One More Time', lyrics by Max Martin.

p.136: 'I'm the hunter' from 'Hunter', lyrics by Björk and Mark Bell.

p.150: 'Every cloud must have a silver lining' from 'My Melancholy Baby', lyrics by George A. Norton.

Picture Credits

The photo of Thomas Edison comes from Library of Congress, Prints&Photographs Division, http://www.loc.gov/pictures/resource/cwpbh.04044/.

The original photo of the Piegan Indian comes from the Herbert E. French photograph collection, Library of Congress, http://www.loc.gov/pictures/item/npc2008000561/. The cleaned-up version is by Harris & Ewing, https://commons.wikimedia.org/w/index.php?curid=6338449.

The human hearing system drawing is based on a vector file from Inductiveload, https://commons.wikimedia.org/w/index.php?curid=5958172, that is a tracing of a figure from Chittka, L., Brockmann, A., 2005. 'Perception Space – The Final Frontier'. *PLOS Biology*, 3(4): e137. https://doi.org/10.1371/journal.pbio.0030137.

The vocal tract measured in an MRI scanner is reproduced with permission from Daniel Aalto. Aalto, D., Aaltonen, O., Happonen, R.P., Jääsaari, P., Kivelä, A., Kuortti, J., Luukinen, J.M., Malinen, J., Murtola, T., Parkkola, R. and Saunavaara, J., 2014. 'Large scale data acquisition of simultaneous MRI and speech'. *Applied Acoustics*, 83, pp. 64–75.

The phrenology picture is from the Wellcome Library, London. Chart from *The Phrenological Journal* ('Know Thyself'), print from Dr E. Clark.

A modern view of the brain showing the emphasis on connectivity is courtesy of the Laboratory of Neuro Imaging and Martinos Center for Biomedical Imaging, Consortium of the Human Connectome Project – www.humanconnectomeproject.org.

The cartoon of an opera performance is an engraving by John Vanderbank of *Handel's Flavio*, available in the public domain.

Photos from Pear's experiment taken from Pear, T. H., 1931. *Voice and Personality*. Chapman & Hall, p. 151.

The scone map of the British Isles courtesy of Adrian Leemann, David Britain and Tam Blaxter.

The photo of the dummy head is by dummy head Thorsten Krienke, https://www.flickr.com/photos/krienke/.

The Kempelen machine drawing is from Kempelen, W. von., 1791. *Mechanismus der menschlichen Sprache*. Degen, p. 438. The photo of the replica is courtesy of Fabian Brackhane and Jürgen Trouvain.

The photo of the Voder at the World Fair is from the New York Public Library, catalog ID (B-number): b11686556.

The picture of Android Repliee Q2 is by Max Braun, https://www.flickr.com/photos/maxbraun/.

The photo from *Spillikin: A Love Story* is by Steve Tanner, courtesy of Pipeline Theatre.

The photo of the lie-detector test is by Ed Westcott and is from the US Department of Energy Photo Archives.

Photos of diffusers and applications reproduced from Cox, T. J. and D'Antonio, P., 2016. *Acoustic Absorbers and Diffusers: Theory, Design and Application*. CRC Press.

Notes

Introduction

1 Quote is by Professor Abel FRS, offering thanks having just chaired the sixty--third ordinary general meeting of the Society of Telegraph Engineers on 13 February 1878. *Journal of the Society of Telegraph Engineers*, 7(21), pp. 68–74.

2 Quote comes from a letter from Alfred Mayer, professor of physics at the Stevens Institute, to Thomas Edison, after a demonstration of the phonograph. http://edison.rutgers.edu/yearofinno/TAEBdocs/Doc1175_MayertoTAE_1-15-78.pdf?DocId=D7829C, accessed 13 May 2017.

3 Albert Speer, the minister for armaments. From Huxley, A., 1958. *Brave New World Revisited*. New York and Evanston: Perennial Library.

4 *London Weekly Graphic*, 16 March 1878. 'The Phonograph at the Royal Institution'. Edison's phonograph initially met with scepticism by some, because it seemed too simple to be plausible. A news article about the phonograph prompted an incredulous college professor to accuse the journalist of being 'a common penny-a-liner in the incipient stages of delirium tremens'. Another correspondent considered Edison 'a fool, or a damned scoundrel, or both'. Any doubts were put to rest by the demonstrations that astounded and delighted audiences.

5 This phonograph looks identical to the one I saw at the Royal Institution. It is different to the one used for the first UK demonstration, because the

machine sent by Edison did not arrive in England soon enough. Preece, W. H., 1878. 'The phonograph'. *Journal of the Society of Telegraph Engineers*, 7(21), pp. 68–74.

6 Rubery, M., 2014. 'Thomas Edison's Poetry Machine'. *19: Interdisciplinary Studies in the Long Nineteenth Century.*

7 Edison, T. A., 1878. 'The phonograph and its future'. *North American Review*, 126(262), pp. 527–36.

8 There is great similarity to the plot of 'Be Right Back', a 2013 episode from the drama *Black Mirror.*

9 https://www.theverge.com/a/luka-artificial-intelligence-memorial-roman-Roman-bot, accessed 13 May 2017.

10 https://www.bloomberg.com/news/articles/2016–12–13/why-google-micro-soft-and-amazon-love-the-sound-of-your-voice, accessed 13 May 2017.

11 Thompson, E., 2002. *The Soundscape of Modernity. Architectural Acoustics and the Culture of Listening in America.* MIT Press, p. 49.

12 Pogue, E., 2014. 'Unsettled Score'. *Scientific American.* https://www.scientific-american.com/article/why-digital-music-looks-set-to-replace-live-perfor-mances/, accessed 19 November 2017.

Chapter 1

1 Müller, F. M. (1861) quoted in Noiré, L., 1917. *The origin and philosophy of language.* The Open Court Publishing Company, p. 73. The quote in Müller's Lectures on the Science of Language is slightly different.

2 This overlooks sign language.

3 For more about Darwin's theories and language evolution see Fitch, W. T., 2010. *The Evolution of Language.* Cambridge University Press.

4 Ball, P., 2010. *The Music Instinct: How Music Works and Why We Can't Do Without It.* Random House.

5 Lieberman, D., 2011. *The Evolution of the Human Head.* Harvard University Press.

6 http://museumvictoria.com.au/melbournemuseum/discoverycentre/600–million-years/timeline/devonian/acanthostega/, accessed 21 March 2016.

7 Christensen, C. B., Lauridsen, H., Christensen-Dalsgaard, J., Pedersen, M. and Madsen, P. T., 2015. 'Better than fish on land? Hearing across meta-morphosis in salamanders'. Proceedings of the Royal Society of London *B: Biological Sciences*, 282(1802), p. 20141943.

8 Kitazawa, T., Takechi, M., Hirasawa, T., Adachi, N., Narboux-Nême, N., Kume, H., Maeda, K., Hirai, T., Miyagawa-Tomita, S., Kurihara, Y. and Hitomi, J., 2015. 'Developmental genetic bases behind the independent origin of the tympanic membrane in mammals and diapsids'. *Nature communications*, 6.

9 Yost, W. A., 1994. *Fundamentals of Hearing: An introduction*. Academic Press.

10 The historical quotes concerning Reichert in this section come from Asher, R. J., 2012. 'Evolutionary Biology and Scepticism: the Reception of Darwinism in 19th Century German Embryology' in Calne R. and O'Reilly, W. (eds.), *Scepticism: Hero and Villain*. NOVA publishers, pp. 71–86.

11 This quote may appear bombastic, but that was the style of Haeckel, whose opinions are now overshadowed by his scientific misdemeanours. Vera Weisbecker was very blunt and described him as a 'complete nutcase', recounting how Haeckel's 'bizarre falsification' of data still creates problems for evolutionary biologists today.

12 Grothe, B. and Pecka, M., 2015. 'The natural history of sound localization in mammals – a story of neuronal inhibition'. *Inhibitory Function in Auditory Processing*. This dates the tympanic middle to 210–230 million years ago. This is in the Triassic Period, when the middle ear formed independently in amphibians, reptiles/birds and mammals.

13 Luo, Z. X., Chen, P., Li, G. and Chen, M., 2007. 'A new eutriconodont mammal and evolutionary development in early mammals'. *Nature, 446*(7133), pp. 288–93.

14 Walsh, S. A., Luo, Z. X. and Barrett, P. M., 2013. 'Modern imaging techniques as a window to prehistoric auditory worlds' in *Insights from Comparative Hearing Research*. Springer New York, pp. 227–61.

15 The ability to pick out sounds among a background hubbub is another important driver; see Fay, R. R. and Popper, A. N., 2000. 'Evolution of hearing in vertebrates: the inner ears and processing'. *Hearing Research*, 149(1), pp. 1–10.

16 These bandwidths arise because of the way the size of the sound wave changes with frequency. At low frequency the sound *wavelength* is larger than the head, and so the sound readily bends around to the furthest ear and timing cues are useful. At high frequency the sound wavelength is smaller than the head, and the sound does not bend so easily around to the furthest ear and so level cues are most useful.

17 Martin, T., Marugán-Lobón, J., Vullo, R., Martín-Abad, H., Luo, Z. X. and Buscalioni, A. D., 2015. A Cretaceous eutriconodont and integument evolution in early mammals. *Nature, 526*(7573), pp. 380–4.

18 Quam, R., Martínez, I., Rosa, M., Bonmatí, A., Lorenzo, C., de Ruiter, D. J., Moggi-Cecchi, J., Valverde, M. C., Jarabo, P., Menter, C. G. and Thackeray, J. F., 2015. 'Early hominin auditory capacities'. *Science advances*, 1(8), p. e1500355.

19 http://australianmuseum.net.au/australopithecus-africanus, accessed 21 March 2016. The specimen was found in 1924.

20 While modern humans have lost a little sensitivity in this bandwidth we have a better ability to pick up higher frequency sound.

21 Martínez, I., Rosa, M., Quam, R., Jarabo, P., Lorenzo, C., Bonmatí, A., Gómez-Olivencia, A., Gracia, A. and Arsuaga, J. L., 2013. 'Communicative capacities in Middle Pleistocene humans from the Sierra de Atapuerca in Spain'. *Quaternary International*, 295, pp. 94–101.

22 http://humanorigins.si.edu/evidence/human-fossils/species/homo-heidelbergensis, accessed 21 March 2016. See also Buck, L. T. and Stringer, C. B., 2014. '*Homo heidelbergensis*'. *Current Biology*, 24(6), pp. R214–R215.

23 Very new evidence suggest that *Homo sapiens* might be 100,000 years older than this; Richter, D., Grün, R., Joannes-Boyau, R., Steele, T. E., Amani, F., Rué, M., Fernandes, P., Raynal, J. P., Geraads, D., Ben-Ncer, A. and Hublin, J. J., 2017. 'The age of the hominin fossils from Jebel Irhoud, Morocco, and the origins of the Middle Stone Age'. *Nature*, 546(7657), pp. 293–6.

24 Stoessel, A., David, R., Gunz, P., Schmidt, T., Spoor, F. and Hublin, J. J., 2016. 'Morphology and function of Neandertal and modern human ear ossicles'. *Proceedings of the National Academy of Sciences*, 113(41), pp. 11489–94.

25 A date of 530,000 years ago is often quoted based on some important Spanish fossils.

26 You might also consider a more extreme comparison of chimpanzees and humans. Although we have been following different evolutionary pathways for 6–7 million years, chimps can hear speech and can be trained to respond to commands spoken by humans. Humans do have improved hearing over a vital bandwidth for intelligible speech, roughly 1,000–5,000 Hz. But at most this is twenty-decibels – the difference in loudness between quiet talking and a raised voice. Coleman, M. N., 2009. 'What do primates hear? A meta-analysis of all known nonhuman primate behavioral audiograms'. *International journal of primatology*, 30(1), pp. 55–91.

27 'Neanderthal'. *Oxford Dictionaries*. Oxford University Press. http://www.oxforddictionaries.com/definition/english/neanderthal, accessed 21 March 2016.

28 Pagel, M., 2016. 'How humans evolved language, and who said what first'. *New Scientist*, 229(3059), pp. 26–9.

29 Bolhuis, J. J., Tattersall, I., Chomsky, N. and Berwick, R. C., 2014. 'How could language have evolved?'. *PLOS Biology*, 12(8), p. e1001934.

30 The vocal fold spectrum is an effective frequency response allowing for radiation impedance.

31 Aalto, D., Aaltonen, O., Happonen, R. P., Jääsaari, P., Kivelä, A., Kuortti, J., Luukinen, J. M., Malinen, J., Murtola, T., Parkkola, R. and Saunavaara, J., 2014. 'Large-scale data acquisition of simultaneous MRI and speech'. *Applied Acoustics*, 83, pp. 64–75.

32 One study found that a more monotone voice was a predictor of the number of past sexual partners. Hodges-Simeon, C. R., Gaulin, S. J. and Puts, D. A., 2011. 'Voice correlates of mating success in men: examining "contests" versus "mate choice" modes of sexual selection'. *Archives of Sexual Behavior*, 40(3), pp. 551–7.

33 This is a good example of the development of an individual following what happened in evolution. Fitch, W. T., 2000. 'The evolution of speech: a comparative review'. *Trends in Cognitive Sciences*, 4(7), pp. 258–67.

34 D'Anastasio, R., Wroe, S., Tuniz, C., Mancini, L., Cesana, D. T., Dreossi, D., Ravichandiran, M., Attard, M., Parr, W. C., Agur, A. and Capasso, L., 2013. 'Micro-biomechanics of the Kebara 2 hyoid and its implications for speech in Neanderthals'. *PLOS One*, 8(12), p. e82261.

35 Lieberman, op. cit.

36 Fitch, W. T., de Boer, B., Mathur, N. and Ghazanfar, A. A., 2016. Monkey vocal tracts are speech-ready. Science Advances, 2(12), p. e1600723.

37 Fitch, *The Evolution of Language*, op. cit.

38 Bowling, D. L., Garcia, M., Dunn, J. C., Ruprecht, R., Stewart, A., Frommolt, K. H. and Fitch, W. T., 2017. 'Body size and vocalization in primates and carnivores'. *Scientific Reports*, 7. Lowering the larynx is not the only way of aurally exaggerating body size. On Kenya's Rusinga Island, a fossil of an extinct wildebeest has a greatly enlarged nasal passage that is thought to have enabled a low trumpeting sound. http://news.nationalgeographic.com/2016/02/160204-ancient-wildebeest-fossil-ice-age-dinosaur/, accessed 21 March 2016.

39 Fitch, W. T. and Giedd, J., 1999. 'Morphology and development of the human vocal tract: A study using magnetic resonance imaging'. *Journal of the Acoustical Society of America*, 106(3), pp. 1511–22.

40 Creating more sound at lower frequency also allows calls to carry through the forest.

41 De Boer, B., 2012. 'Loss of air sacs improved hominin speech abilities'. *Journal of Human Evolution*, 62(1), pp. 1–6.

42 At one time it was thought that examining fossils to estimate the size of nerves that control muscles important for speech would be useful, but that now looks unlikely. See for example, Meyer, M. R. and Haeusler, M., 2015. 'Spinal cord evolution in early *Homo*'. *Journal of Human Evolution*, 88, pp. 43–53.

43 Lieberman, op. cit. Also Schoenemann, P. T., 2006. 'Evolution of the size and functional areas of the human brain'. *Annual Review of Anthropology*, 35, pp. 379–406.

44 Lieberman, op. cit.

45 People usually name sources (e.g. rain) rather than use an onomatopoeia thet describes the sound (e.g. pitter-patter). Bones, O. C., Davies, W. J. and Cox, T. J., 2017. 'Clang, chitter, crunch: Perceptual organisation of onomatopoeia'. *Journal of the Acoustical Society of America*, 141, p. 3694.

46 Otto, J., 1922. *Language; Its Nature, Development and Origin*. G. Allen & Unwin Ltd.

47 Some would argue that the starting point for language evolution would be words rather than whole phrases; syntactic operations would develop afterwards for combining words into sentences. Vocal imitation also played a central role in allowing a large vocabulary to develop.

48 These words start to emerge because of chance associations that happen to be in the random strings of text early on in the experiment. Kirby, S., Cornish, H. and Smith, K., 2008. 'Cumulative cultural evolution in the laboratory: An experimental approach to the origins of structure in human language'. *Proceedings of the National Academy of Sciences*, 105(31), pp. 10681–6.

49 Simon has recently been researching languages that are spontaneously appearing. This happens within isolated communities where deaf children start being born because of hereditary conditions. Parents and children have to invent a sign language to allow communication.

50 Bolhuis et al., op. cit.

51 http://humanorigins.si.edu/evidence/human-fossils/fossils/la-chapelle-aux-saints, accessed 1 April 2017.

52 Wells, H. G., 1921. *The Outline of History: Volume 1*. Macmillan, New York, p. 67.

53 The exception is for those from sub-Saharan Africa. Fu, Q., Hajdinjak, M., Moldovan, O. T., Constantin, S., Mallick, S., Skoglund, P., Patterson, N., Rohland, N., Lazaridis, I., Nickel, B. and Viola, B., 2015. 'An early modern human from Romania with a recent Neanderthal ancestor'. *Nature*, 524(7564), pp. 216–19.

54 http://www.nhm.ac.uk/press-office/press-releases/a-comment-on-_ancient-gene-flow-from-early-modern-humans-into-ea.html#sthash.cvJmZJFI.dpuf, accessed 21 March 2016.

55 Lieberman, op. cit.

56 But the FOXP2 in Neanderthals is not exactly the same as that found in *Homo sapiens*. The genetic data points to mutations forming the modern human FOXP2 1.8–1.9 million years ago.

Chapter 2

1 Mehl, M. R., Vazire, S., Ramírez-Esparza, N., Slatcher, R. B. and Pennebaker, J. W., 2007. 'Are women really more talkative than men?' *Science*, 317(5834), p. 82. According to this paper we speak an average of 16,000 words per day, so we average 6 million words per year. With current life expectancy being eighty years that gives near 500 million words in a lifetime.

2 Wolke, D., Bilgin, A. and Samara, M., 2017. 'Systematic Review and Meta-Analysis: Fussing and Crying Durations and Prevalence of Colic in Infants'. *Journal of Pediatrics*, DOI: 10.1016/j.jpeds.2017.02.020.

3 Wermke, K. and Mende, W., 2011. 'From emotion to notion: the importance of melody' in Decety, J. and Cacioppo, J. T. *The Oxford Handbook of Social Neuroscience*. Oxford University Press, USA.

4 Mampe, B., Friederici, A. D., Christophe, A. and Wermke, K., 2009. 'Newborns' cry melody is shaped by their native language'. *Current Biology*, 19(23), pp. 1994–7.

5 Quite a few studies are exploring the effectiveness of playing maternal womb sounds to premature babies, including Rand, K. and Lahav, A., 2014. 'Maternal sounds elicit lower heart rate in preterm newborns in the first month of life'. *Early human development*, 90(10), pp. 679–83.

6 *Caveat emptor*; this paper involves a manufacturer of a product. López-Teijón, M., García-Faura, Á. and Prats-Galino, A., 2015. 'Fetal facial expression in response to intravaginal music emission'. *Ultrasound*, 23(4), pp. 216–23.

7 For ethical reasons no one has examined a baby's vocal folds to confirm what is going on.

8 This information is based on studying 100,000 cries. Wermke, op. cit.

9 http://www.wired.com/2009/11/iphone-application-translates-babies-howls/, accessed 28 July 2016.

10 Kuhl, P. K., 2015. 'Baby Talk'. *Scientific American*, 313(5), pp. 64–9.

11 Vouloumanos, A., Hauser, M. D., Werker, J. F. and Martin, A., 2010. 'The tuning of human neonates' preference for speech'. *Child development*, 81(2), pp. 517–27.

12 The voices of the Clangers were made on swanee or slide whistles.

13 Another test compared forward and backwards speech: Pena, M., Maki, A., Kovačić, D., Dehaene-Lambertz, G., Koizumi, H., Bouquet, F. and Mehler, J., 2003. 'Sounds and silence: an optical topography study of language recognition at birth'. *Proceedings of the National Academy of Sciences*, 100(20), pp. 11702–5.

14 Graddol, D. and Swann, J., 1983. 'Speaking fundamental frequency: some physical and social correlates'. *Language and Speech*, 26(4), pp. 351–66.

15 Kuhl, op. cit.

16 'Robot companion's can-do attitude rubs off on children'. http://www.newscientist.com /article/mg23331144-100-robot-companions-cando-attitude-rubs-off-on-children/, accessed 3 April 2017.

17 Roy, B. C., Frank, M. C., DeCamp, P., Miller, M. and Roy, D., 2015. 'Predicting the birth of a spoken word'. *Proceedings of the National Academy of Sciences*, 112(41), pp. 12663–8.

18 This is transcribed from Roy's TED talk. http://www.ted.com/talks/deb_roy_the_birth_of_a_word?language=en, accessed 28 July 2016.

19 Ramírez-Esparza, N., García-Sierra, A. and Kuhl, P. K., 2014. 'Look who's talking: speech style and social context in language input to infants are linked to concurrent and future speech development'. *Developmental Science*, 17(6), pp. 880–91.

20 Curtiss, S., 2014. 'Genie: a psycholinguistic study of a modern-day wild child'. Academic Press.

21 Similar comments could also be made about the plight of Romanian orphans, with various studies showing that institutionalisation with limited language and social input affects language acquisition.

22 https://www.ted.com/talks/patricia_kuhl_the_linguistic_genius_of_babies?language=en, accessed 28 July 2016.

23 Hakuta, K., Bialystok, E. and Wiley, E., 2003. 'Critical evidence a test of the critical-period hypothesis for second-language acquisition'. *Psychological Science*, 14(1), pp. 31–8.
24 Scovel, T., 2000. 'A critical review of the critical period research'. *Annual Review of Applied Linguistics*, 20, pp. 213–23.
25 For one in five left-handed people, the language processing is focused on the right-hand side. http://www.rightleftrightwrong.com/brain.html, accessed 17 May 2017.
26 Plaza, M., Gatignol, P., Leroy, M. and Duffau, H., 2009. 'Speaking without Broca's area after tumor resection'. *Neurocase*, 15(4), pp. 294–310.
27 Miller, N., 2016. 'Stuttering isn't only psychological – and a cure might be coming'. *New Scientist*, 3067.
28 Careful experimental design is needed, however, because there are difficulties with using fMRI for stammering studies. Turning the scanner's powerful magnets on and off creates huge forces in the electromagnetic coils and thus creates extremely loud rhythmic noise. Listening to this can aid fluency!
29 While clear differences can be seen between the brains of those who stammer and those who don't, we do not know if these structural differences cause the stammering or whether they are a consequence of brain development inhibited by stuttering.
30 I chatted to Patrick over lunch at the conference, but this quote is taken from the BSA blog https://www.stammering.org/what-we-do/blog/what-if-we-fight-our-right-stammer, accessed 29 July 2016.
31 https://www.britannica.com/biography/Lewis-Carroll, accessed 29 July 2016.
32 http://www.stutteringhelp.org/famous-people/lewis-carroll, accessed 29 July 2016.
33 A good example of this environmental influence is that before puberty boys' voices tend to have lower formants than girls' even though the vocal anatomy of both genders is similar. By making small adjustments to the length of the vocal tract, the children feminise or masculinise their voices.
34 Schneider, B. and Bigenzahn, W., 2003. 'Influence of glottal closure configuration on vocal efficacy in young normal-speaking women'. *Journal of Voice*, 17(4), pp. 468–80.
35 Xu, Y., Lee, A., Wu, W.L., Liu, X. and Birkholz, P., 2013. 'Human vocal attractiveness as signaled by body size projection'. *PLOS One, 8(*4), p. e62397.

36 Attractive male voices are also more monotone. Puts, D. A., 2005. 'Mating context and menstrual phase affect women's preferences for male voice pitch'. *Evolution and Human Behavior*, 26(5), pp. 388–97.

37 Scott, S. and McGettigan, C., 2016. 'The voice: From identity to interactions' in Matsumoto, D., Hwang, H. C. and Frank, M. G. (eds). 'APA handbook of nonverbal communication'. *American Psychological Association*, pp. 289–305.

38 Female voices also vary over the menstrual cycle and offer a weak cue to when the female is most fertile. Fischer, J., Semple, S., Fickenscher, G., Jürgens, R., Kruse, E., Heistermann, M. and Amir, O., 2011. 'Do women's voices provide cues of the likelihood of ovulation? The importance of sampling regime'. *PLOS One*, 6(9), p. e24490.

39 Simmons, L. W., Peters, M. and Rhodes, G., 2011. 'Low-pitched voices are perceived as masculine and attractive but do they predict semen quality in men?'. *PLOS One*, 6(12), p. e29271.

40 Hatzinger, M., Vöge, D., Stastny, M., Moll, F. and Sohn, M., 2012. 'Castrati singers – All for fame'. *Journal of Sexual Medicine*, 9(9), pp. 2233–7.

41 https://www.theguardian.com/music/2002/aug/05/classicalmusicandopera.artsfeatures, accessed 29 July 2016.

42 https://youtu.be/KLjvfqnD0ws, accessed 29 July 2016. Some argue that Moreschi's voice had passed its best when the recordings were made.

43 He recorded the sound of a boy soprano, and then processed it in a computer to remove the resonances of the vocal tract. What was left was an approximation of what the boy's vocal folds were creating. A computer simulation of the resonances created by a baritone's vocal tract was then used to augment the sound of the boy's vocal folds, boosting some frequencies and attenuating others. Thus the simulation assumed that the castrato had a vocal tract like an adult male.

44 Maybe the castrati used their vocal tract resonances like a modern soprano rather than a baritone? Chapter 5 has more on how modern male and female opera singers use their formants differently.

45 Quote from Jenkins, J. S., 1998. 'The voice of the castrato'. *Lancet*, 351(9119), pp. 1877–80, where it is credted to De Brosses, C. (1799). *Lettres historiques et critiques sur l'Italie*. 3 vols., Paris, 3, p. 246.

46 http://www.telegraph.co.uk/news/health/3312210/Are-you-damaging-your-voice.html, accessed 19 August 2017. The charity Voice Care Network provides valuable advice on how teachers can maintain a healthy voice: http://voicecare.org.uk/.

47 We speak 6 million words a year (see note 1) and on average words last about 0.3 seconds. Yuan, J., Liberman, M. and Cieri, C., September 2006. 'Towards an integrated understanding of speaking rate in conversation'. *Interspeech*. This means we are talking for about 2 million seconds a year (which is equivalent to twenty-four days!). Assuming a male talker at a pitch of 120 Hz, this means the vocal folds open and close a little over 200 million times every year.

48 The lungs become less efficient and their capacity typically drops by about 40 per cent.

49 The change in pitch on Cooke's voice is hard to spot in later broadcasts because the producers boosted the bass frequencies. Analysis comes from Reubold, U., Harrington, J. and Kleber, F., 2010. 'Vocal aging effects on F 0 and the first formant: a longitudinal analysis in adult speakers'. *Speech Communication*, 52(7), pp. 638–51.

50 Pemberton, C., McCormack, P. and Russell, A., 1998. 'Have women's voices lowered across time? A cross-sectional study of Australian women's voices'. *Journal of Voice*, 12(2), pp. 208–13.

51 https://www.theguardian.com/music/2007/jul/05/popandrock1, accessed 29 July 2016.

52 For more on expectation in music, see Ball, op. cit.

53 http://www.parliament.uk/business/publications/research/key-issues-for-the-new-parliament/value-for-money-in-public-services/the-ageing-population/, accessed 29 July 2016.

54 Golub, J. S., Chen, P. H., Otto, K. J., Hapner, E. and Johns, M. M., 2006. 'Prevalence of perceived dysphonia in a geriatric population'. *Journal of the American Geriatrics Society*, 54(11), pp. 1736–9. See also Johns, M. M., Arviso, L. C. and Ramadan, F., 2011. 'Challenges and opportunities in the management of the aging voice'. *Otolaryngology – Head and Neck Surgery*.

55 See 'Voice lifts: something to shout about', *Guardian*, https://www.theguardian.com/lifeandstyle/2012/sep/23/voice-lift-vocal-cord-treatment, and '"Voice Lift" Surgery, In Most Cases, Not Worth It', http://www.seattleplasticsurgery.com/blog/2012/11/%E2%80%9Cvoice-lift%E2%80%9D-surgery-in-most-cases-not-worth-it/, both accessed 7 February 2017.

56 See http://voicecare.org.uk/ for other advice on maintaining vocal health.

57 Tay, E. Y. L., Phyland, D. J. and Oates, J., 2012. 'The effect of vocal function exercises on the voices of aging community choral singers'. *Journal of Voice*, 26(5), pp. 672–e19.

58 Stemple, J. C., 2002. *Vocal function exercises*. Plural Publishing Incorporated.
59 Prakup, B., 2012. 'Acoustic measures of the voices of older singers and nonsingers'. *Journal of Voice*, 26(3), pp. 341–50.
60 Lortie, C. L., Rivard, J., Thibeault, M. and Tremblay, P., 2016. 'The Moderating Effect of Frequent Singing on Voice Aging'. *Journal of Voice*, 31(1), pp. 112.e1–e12.

Chapter 3

1 Monrad-Krohn, G. H., 1947. 'Dysprosody or altered "melody of language"'. *Brain: a journal of neurology.*
2 The interviews were done at least six months after the condition began. Miller, N., Taylor, J., Howe, C. and Read, J., 2011. 'Living with foreign accent syndrome: insider perspectives'. *Aphasiology*, 25(9), pp. 1053–68.
3 DiLollo, A., Scherz, J. and Neimeyer, R. A., 2014. 'Psychosocial implications of foreign accent syndrome: two case examples'. *Journal of Constructivist Psychology*, 27(1), pp. 14–30.
4 Some cases also talk about relatives showing anger, fear and disbelief, because they assume that the voice is being put on by the patient.
5 David Thorpe's documentary film *Do I Sound Gay?* (2014) explores this issue, including examining why some gay males turn to speech therapists. This autobiographical film shows that while some perceive their voice to be a problem, often this is symptomatic of deeper psychological issues.
6 Rule, N. O., 2017. 'Perceptions of sexual orientation from minimal cues'. *Archives of Sexual Behavior*, 46(1), pp. 129–39.
7 A study compared actors who played both straight and gay characters. They found that when the actors were playing homosexual characters they raised the pitch of their voice towards the top of the typical male range. They also found the actors made the gay voices more melodious by having greater pitch variation. Cartei, V. and Reby, D., 2012. 'Acting gay: Male actors shift the frequency components of their voices towards female values when playing homosexual characters'. *Journal of Nonverbal Behavior*, 36(1), pp. 79–93.
8 Other features less studied but commonly associated with a stereotypical gay voice are clearer and longer vowels, 'l's being clearer and over-articulated 'p's, 't's and 'k's. See *Do I Sound Gay?*. Also 'vocal fry' (a creaking sound like the one Britney Spears uses in the opening of '… Baby One More Time') and sentences that end with a rising intonation.

9 Van Borsel, J. and Van de Putte, A., 2014. 'Lisping and male homosexuality'. *Archives of Sexual Behavior*, 43(6), pp. 1159–63.

10 Another example is that male voice pitch tends to be towards the bottom end of what their vocal anatomy can achieve. Something that is not true of females. It is assumed that men are driven to use the lower part of the vocal range because that is found to be more attractive. Graddol, D. and Swann, J., 1983. 'Speaking fundamental frequency: some physical and social correlates'. *Language and Speech*, 26(4), pp. 351–66.

11 In the past, the stereotype was reinforced by the only gay vocal role models in the media being camp celebrities like Larry Grayson, the host of *The Generation Game* and *Blankety Blank*.

12 There are thought to be many more who have not come forward for medical advice. One estimate is that 0.2 per cent of people have sufficient incongruity with their gender to consider medical intervention. https://www.theguardian.com/society/2016/jul/10/transgender-clinic-waiting-times-patient-numbers-soar-gender-identity-services, accessed 29 July 2016.

13 The literature gives a confusing picture because voice transition requires speech therapy alongside the surgery to gain a feminine voice, and the quality and quantity of the therapy varies considerably.

14 Hancock, A. and Helenius, L., 2012. 'Adolescent male-to-female transgender voice and communication therapy'. *Journal of Communication Disorders*, 45(5), pp. 313–24.

15 Davies, S., Papp, V. G. and Antoni, C., 2015. 'Voice and communication change for gender nonconforming individuals: giving voice to the person inside'. *International Journal of Transgenderism*, 16(3), pp. 117–59.

16 Hillenbrand, J. M. and Clark, M. J., 2009. 'The role of f 0 and formant frequencies in distinguishing the voices of men and women'. *Attention, Perception, & Psychophysics*, 71(5), pp. 1150–66.

17 This is based on comments by Christella. A small number of scientific studies have looked at the success of changing these features but results are inconclusive.

18 It is alleged that Pythagoras' followers, the Acousmatics, were only allowed to hear their master's disembodied voice from behind a curtain. But some have cast doubt on this. Kane, B., 2014. *Sound Unseen: Acousmatic Sound in Theory and Practice*. Oxford University Press, USA.

19 Pear, T. H., 1931. *Voice and Personality*. Chapman & Hall, p. 151. The quotes in this section all come from this book.

20 Lehr, S. and Banaji, M., 2011. 'Implicit Association Test (IAT)'. *Oxford Bibliographies* in Psychology. doi: 10.1093/obo/9780199828340-0033.

21 Kalat, J. W., 2015. *Biological psychology.* Nelson Education.

22 Johnson, D. R., Cushman, G. K., Borden, L. A. and McCune, M. S., 2013. 'Potentiating empathic growth: Generating imagery while reading fiction increases empathy and prosocial behavior'. *Psychology of Aesthetics, Creativity, and the Arts*, 7(3), p. 306.

23 Brain studies reveal that as readers conjure more vivid imagery, more brain regions are active alongside the 'classic' language centres of the mind. To take one example, when we interpret certain metaphors, we co-opt sensory and motor parts of the brain. If you read about someone having a 'rough day', or a 'slimy person', then sensory regions of the brain linked to touch become engaged. Lacey, S., Stilla, R. and Sathian, K., 2012. 'Metaphorically feeling: comprehending textural metaphors activates somatosensory cortex'. *Brain and Language*, 120(3), pp. 416–21.

24 Andersen, E. S., 1984. 'The acquisition of sociolinguistic knowledge: Some evidence from children's verbal role-play'. *Western Journal of Communication (includes Communication Reports)*, 48(2), pp. 125–44.

25 Kreiman, J. and Sidtis, D., 2011. *Foundations of voice studies: An interdisciplinary approach to voice production and perception.* John Wiley & Sons.

26 Listeners tend to overestimate the age of young adults. This probably arises because of a *floor effect*; for someone with an adult voice there is an age below which respondents will not suggest, say fifteen, but there are plenty of older ages that can be guessed.

27 If you need to estimate height, it is better to listen for vocal timbre created by resonances of the airways. You can then estimate height to within about 10 cm. Morton, J., Sommers, M., Lulich, S., Alwan, A. and Arsikere, H., 2013. 'Acoustic features mediating height estimation from human speech'. *Journal of the Acoustical Society of America*, 134(5), p. 4072.

28 This is the current favoured model for recognition, but this is not universally agreed upon. Mathias, S. R. and von Kriegstein, K., 2014. 'How do we recognise who is speaking?'. *Front Biosci (Schol Ed)*, 6, pp. 92–109.

29 https://www.judiciary.gov.uk/wp-content/uploads/2014/12/r-v-dwaine-george.pdf, accessed 8 October 2016.

30 Legge, G. E., Grosmann, C. and Pieper, C. M., 1984. 'Learning unfamiliar voices'. *Journal of Experimental Psychology: Learning, Memory, and Cognition*, 10(2), p. 298.

31 Saslove, H. and Yarmey, A. D., 1980. 'Long-term auditory memory: Speaker identification'. *Journal of Applied Psychology*, 65(1), p. 111.
32 Studies show recognition from 'hello' with familiar voices has a 20–60 per cent success rate. There is a big variation because it depends greatly on experimental design. Kreiman and Sidtis, op. cit., p. 177.
33 Wolfe, T., 1987. *The Bonfire of the Vanities*. Vintage, pp. 16–17.
34 Kreiman and Sidtis, op. cit., pp. 160–2.
35 Penguins are played a processed version of their own call, for example the pitch might be altered. For nesting penguins like the gentoo, if you change the pitch too much the penguins no longer recognise the call. Playing the call backwards, on the other hand, has no effect on recognition. This shows that timing and rhythm are not so important.
36 Kreiman and Sidtis, op. cit., p. 182.
37 http://whatsnext.nuance.com/customer-experience/five-common-voice-biometrics-myths/, accessed 8 October 2016.
38 'First Impressions'. *Wired*, May 2016.
39 http://www.bbc.co.uk/news/technology-39965545, accessed 1 June 2017.
40 Waugh, P., 2015. 'The novelist as voice hearer'. *Lancet*, 386(10010), pp. e54–e55. I also liked the description in this paper, of how writers 'harness the power of the inner voice to create imaginary characters whose thoughts and feelings entangle with those of real readers'.
41 Perrone-Bertolotti, M., Rapin, L., Lachaux, J. P., Baciu, M. and Loevenbruck, H., 2014. 'What is that little voice inside my head? Inner speech phenomenology, its role in cognitive performance, and its relation to self-monitoring'. *Behavioural Brain Research*, 261, pp. 220–39.
42 Initially, the silence actually created an almost euphoric state, the quote continues – 'But then I was immediately captivated by the magnificence of the energy around me. And because I could no longer identify the boundaries of my body, I felt enormous and expansive. I felt at one with all the energy that was, and it was beautiful there.' http://www.ted.com/talks/jill_bolte_taylor_s_powerful_stroke_of_insight/transcript?language=en, accessed 17 May 2017. See also Morin, A., 2009. 'Self-awareness deficits following loss of inner speech: Dr. Jill Bolte Taylor's case study'. *Consciousness and Cognition*, 18(2), pp. 524–9.
43 Filik, R. and Barber, E., 2011. 'Inner speech during silent reading reflects the reader's regional accent'. *PLOS One*, *6*(10), p. e25782.
44 To give another example, people with a stammer are often more fluent with inner speech.

45 Perrone-Bertolotti et al., op. cit.

46 Woods, A., Jones, N., Alderson-Day, B., Callard, F. and Fernyhough, C., 2015. 'Experiences of hearing voices: analysis of a novel phenomenological survey'. *Lancet Psychiatry*, 2(4), pp. 323–31.

47 Wilkinson, S. and Bell, V., 2016. 'The representation of agents in auditory verbal hallucinations'. *Mind & Language*, 31(1), pp. 104–26.

48 Alderson-Day, B., Bernini, M. and Fernyhough, C., 2017. 'Uncharted features and dynamics of reading: Voices, characters, and crossing of experiences'. *Consciousness and Cognition*, 49, pp. 98–109.

Chapter 4

1 Data from the campaign up until 1 July 2016. https://www.washingtonpost.com/news/the-fix/wp/2016/07/01/donald-trump-has-been-wrong-way-more-often-than-all-the-other-2016-candidates-combined/, accessed 21 November 2016.

2 Atwill, J. M., 2009. *Rhetoric reclaimed: Aristotle and the liberal arts tradition.* Cornell University Press, p. 37.

3 Of course this attempt to change the voice did not work. Hillary Clinton's accent has also came under intense scrutiny, with, for example, commentators lampooning her use of a Southern drawl when campaigning in the South. http://www.nytimes.com/2015/03/21/us/politics/scott-walker-hones-his-image-among-republicans-for-possible-presidential-race.html?_r=0, accessed 21 November 2016

4 Crystal, B. and Crystal D., 2015. *You Say Potato: The Story of English Accents.* Macmillan, p. 63. Also see www.bl.uk/learning/langlit/sounds/find-out-more/received-pronunciation/, accessed 21 November 2016.

5 For those who like to complain about American actors delivering Shakespeare, a modern American accent is probably closer to what was heard in a Shakespeare play than any current English accent. Crystal and Crystal, op. cit.

6 This tradition continued largely unchallenged for much of the twentieth century, except for a short experiment using regional accents during the Second World War.

7 Ironically, Cary Grant was actually born and brought up in Bristol, England. If Mason had used a strong regional accent from Britain then American audiences may have struggled to understand it. When *Trainspotting* was released in the US, Ewan McGregor was re-voiced with a softer Scottish accent.

8 The full poem is, 'Five plump peas in a pea pod pressed. One grew, two grew, and so did all the rest. They grew and grew and grew and grew and grew and never stopped. Till they grew so plump and perky that the pea pod popped!'

9 In some countries, such as Switzerland, accent is more tied to geography than social status.

10 Clark, L., 2016. 'Fish "chat" to each other and may have "regional accents"'. *Wired.* http://www.wired.co.uk/article/listening-to-regional-accents-of-cod, accessed 21 November 2016.

11 If one group goes into a different habitat that might force a change to the calls to improve communication. That will then drive a more rapid divergence of vocalisations. For more on the role of accents in human evolution see Cohen, E., 2012. 'The evolution of tag-based cooperation in humans'. *Current Anthropology*, 53(5), pp. 588–616.

12 Reading about iconicity inspired me to do a study on onomatopoeia. Bones, O. C., Davies, W. J. and Cox, T. J., 2017. 'Clang, chitter, crunch: Perceptual organisation of onomatopoeia'. *Journal of the Acoustical Society of America*, 141, p. 3694.

13 Another example of universals comes from a study where two shapes are matched to the made-up words 'bouba' and 'kiki'. People tend to match bouba to a round figure and kiki to a spiky shape. What the 2016 study showed for the first time was that universals are a bit more common than previously assumed. Blasi, D. E., Wichmann, S., Hammarström, H., Stadler, P. F. and Christiansen, M. H., 2016. 'Sound–meaning association biases evidenced across thousands of languages'. *Proceedings of the National Academy of Sciences*, p. 201605782.

14 Kaplan, S., 2016. 'A nose by any other name: Biology may affect the way we invent words'. *Washington Post.* https://www.washingtonpost.com/news/speaking-of-science/wp/2016/09/12/a-nose-by-any-other-name-biology-may-affect-the-way-we-invent-words/, accessed 21 November 2016.

15 One of the reasons is the diversity of the population in the distant past. Tribes from various parts of Europe settled in the British Isles and these groups were sufficiently isolated to allow variations in speech to be maintained. Crystal and Crystal, op. cit.

16 McDonnell, A., 2016. 'It's scone as in "gone" not scone as in "bone"'. YouGov. https://yougov.co.uk/news/2016/10/31/its-scone-gone-not-scone-bone/, accessed 21 November 2016. Here is the breakdown of the numbers: 'bone'

ABC1 40 per cent; 'gone' ABC1 55 per cent; 'bone' C2DE 45 per cent; 'gone' C2DE 46 per cent. These do not add up to 100 per cent because of other/don't know.

17 Another approach to gathering data has enabled the mapping of new terms in American. Based on almost a billion geotagged tweets, Jack Grieve from Aston University has produced maps showing that shit, damn and bitch are more common in the south-east compared to the rest of the US. Gajanan, M., 2015. 'Want to know how to curse like a proper American? Have a look at these maps'. *Guardian*.

18 Also, lexical variants are not often used, especially compared to a common vowel pronunciation. This makes it harder to change the pronunciation because it is so ingrained in your way of speaking.

19 Leeman hypothesises that this was due to the movement away from cities after the Second World War, leading to a dilution of rural accents.

20 It also applies to the pronunciation of 'r' at the ends of words, e.g. 'far'.

21 This is not true of all English speakers: Scotland, Ireland and America still have many who pronounce these 'r's.

22 Quinn, B., 2011. 'David Starkey claims "the whites have become black"'. *Guardian*. Starkey also referenced Enoch Powell's 'rivers of blood' speech.

23 Fox, S., 2015. *The New Cockney: New Ethnicities and Adolescent Speech in the Traditional East End of London*. Palgrave Macmillan.

24 Sue Fox used this quote from Professor Penelope Eckert from Stanford University. It is true for Western industrialised societies.

25 Kerswill, P., 2011. TEDxEastEnd https://www.youtube.com/watch?v=hAnFbJ65KYM, accessed 21 November 2016.

26 Aitchinson, J., 1996. 'Is our language sick?'. *Independent*. http://www.independent.co.uk/life-style/reith-lectures-is-our-language-in-decay-1317695.html, accessed 21 November 2016.

27 In another study, Sue Fox has found that people as young as four are picking up on these linguistic features, which implies it is not just a teenage fad and likely to be a long-term language change.

28 McGlone, M. S. and Tofighbakhsh, J., 1999. 'The Keats heuristic: Rhyme as reason in aphorism interpretation'. *Poetics*, 26(4), pp. 235–44.

29 Guerini, M., Özbal, G. and Strapparava, C., 2015. 'Echoes of persuasion: The effect of euphony in persuasive communication'. arXiv preprint arXiv:1508.05817. Success rate is 72–88 per cent when the system is devised and tested on one dataset, for instance develop used Twitter information and

tested on tweets. If the system is then applied to another database, for example on movie slogans, success rates drop to 50–60 per cent.

30 Plosives are also more common in brand names.

31 Lev-Ari, S. and Keysar, B., 2010. 'Why don't we believe non-native speakers? The influence of accent on credibility'. *Journal of Experimental Social Psychology*, 46(6), pp. 1093–6.

32 The change for heavy accent was 5 per cent along the scale being measured. The experiment was set up so that the person was recounting a statement from a native speaker to try and remove prejudice.

33 http://www.lasvegas.videobooth.tv/, accessed 21 November 2016.

34 Seventy per cent of the applause could be explained by seven rhetorical devices. Heritage, J. and Greatbatch, D., 1986. 'Generating applause: A study of rhetoric and response at party political conferences'. *American Journal of Sociology*, 92(1), pp. 110–57.

35 This is about applause signalled by the speaker, as opposed to spontaneous clapping in unexpected places. Note, in another study the figure was about a third.

36 http://news.bbc.co.uk/1/hi/magazine/8128271.stm, accessed 21 November 2016.

37 Huettel, S. A., Mack, P. B. and McCarthy, G., 2002. 'Perceiving patterns in random series: dynamic processing of sequence in prefrontal cortex'. *Nature Neuroscience*, 5(5), pp. 485–90.

38 Shu, S. B. and Carlson, K. A., 2014. 'When three charms but four alarms: identifying the optimal number of claims in persuasion settings'. *Journal of Marketing*, 78(1), pp. 127–39.

39 Another powerful rhetorical device is metaphor, for example Trump's use of the phrase 'we're going to drain the swamp' in Washington.

40 Bull, P. E., 1986. 'The use of hand gesture in political speeches: Some case studies'. *Journal of Language and Social Psychology*, 5, pp. 103–18.

41 Voice pitch has also been looked at in other professions. CEOs with lower voices get bigger remuneration and lead bigger companies. But this is an association study and so low-pitched voices might not lead to success because causation has not been proved. Mayew, W. J., Parsons, C. A. and Venkatachalam, M., 2013. 'Voice pitch and the labor market success of male chief executive officers'. *Evolution and Human Behavior*, 34(4), pp. 243–8.

42 Atkinson, M., 1984. *Our Masters' Voices: The Language and Body Language of Politics*. Psychology Press, p. 113.

43 The comments from Beard come from Davies, C., 2014, 'Mary Beard: vocal women treated as "freakish androgynes"', *Guardian*, and Dowell, B., 2014, 'Mary Beard suffers "truly vile" online abuse after *Question Time*', *Guardian*.

44 Reeve, E., 2015. 'Why Do So Many People Hate the Sound of Hillary Clinton's Voice?'. *New Republic*. https://newrepublic.com/article/121643/why-do-so-many-people-hate-sound-hillary-clintons-voice, accessed 21 November 2016.

45 While it works for politicians, a study showed that male lawyers rated as speaking with less-masculine voices were more likely to win. One reason for this might be that the lawyers unconsciously adopt a more masculine speaking style when they know they have a weak case.

46 Klofstad, C. A., Nowicki, S. and Anderson, R. C., 2016. 'How Voice Pitch Influences Our Choice of Leaders'. *American Scientist*, 104(5), p. 282.

47 Tigue, C. C., Borak, D. J., O'Connor, J. J., Schandl, C. and Feinberg, D. R., 2012. 'Voice pitch influences voting behavior'. *Evolution and Human Behavior*, 33(3), pp. 210–16.

48 Klofstad, C. A., Anderson, R. C. and Nowicki, S., 2015. 'Perceptions of competence, strength, and age influence voters to select leaders with lower-pitched voices'. *PLOS One*, 10(8), p. e0133779.

49 Klofstad, C. A., 2015. 'Candidate voice pitch influences election outcomes'. *Political Psychology*. There were some exceptions, for example in a contest between male and female candidates a higher-pitched male is better. Klofstad speculates this might be about lower-pitched males coming across as too aggressive in this case.

50 Gupta, R., 2011. 'What is Vocal Fry?'. https://www.ohniww.org/katy-perry-voice-vocal-fry/, accessed 21 November 2016.

51 Anderson, R. C., Klofstad, C. A., Mayew, W. J. and Venkatachalam, M., 2014. 'Vocal fry may undermine the success of young women in the labor market'. *PLOS One*, 9(5), p. e97506.

52 A side effect of this hormone exposure, however, is to suppress immune function. So only otherwise healthy individuals can afford to have it coursing through their bodies at high levels. Voice pitch could therefore be signalling gene quality. This would also explain why people who are physically more symmetrical have voices that are more attractive. Hughes, S. M., Pastizzo, M. J. and Gallup Jr, G. G., 2008. 'The sound of symmetry revisited: Subjective and objective analyses of voice'. *Journal of Nonverbal Behavior*, 32(2), pp. 93–108.

53 Cheng, J. T., Tracy, J. L., Ho, S. and Henrich, J., 2016. 'Listen, follow me: Dynamic vocal signals of dominance predict emergent social rank in humans'. *Journal of experimental psychology: general*, 145(5), p. 536.

54 Rosenberg, A. and Hirschberg, J., 2009. 'Charisma perception from text and speech'. *Speech Communication*, 51(7), pp. 640–55.

55 Where there is strong evidence, slower delivery is better. Von Hippel, W., Ronay, R., Baker, E., Kjelsaas, K. and Murphy, S. C., 2016. 'Quick Thinkers Are Smooth Talkers: Mental Speed Facilitates Charisma'. *Psychological Science*, 27(1), pp. 119–22.

56 Jürgens, R., Grass, A., Drolet, M. and Fischer, J., 2015. 'Effect of Acting Experience on Emotion Expression and Recognition in Voice: Non-Actors Provide Better Stimuli than Expected'. *Journal of Nonverbal Behavior*, 39(3), pp. 195–214.

57 The speech contour might be exaggerated by the acting process of reading out the phrases as people naturally do this to make it sound more engaging. This is certainly something I have to do when presenting on radio.

Chapter 5

1 Milner, G., 2011. *Perfecting Sound Forever: The Story of Recorded Music*. Granta Books.

2 Bing Crosby and Al Jolson. 'Alexander's Ragtime Band'. https://www.youtube.com/watch?v=m4q2v-Gavvg, accessed 18 March 2017. At the start of the recording, there is more accommodation happening between the two styles of singing; see Potter, J., 2006. *Vocal Authority: Singing Style and Ideology*. Cambridge University Press.

3 The agent was Art Klein. Freedland, M., 1985. *Jolie: The Al Jolson Story*. W. H. Allen, p. 52.

4 It is important not to overlook cultural factors. Potter, op. cit. About how audience behaviour has changed, also see Byrne, D., 2012. *How Music Works*. Canongate Books.

5 BBC, 1956. *The Listener*.

6 Another problem was that Welsh cliché: the distant sound of a male voice choir. The producer employed buskers to perform from the corridor outside the studio. By opening and closing the soundproof door to the studio he could then alter how loud and how distant the choir sounded.

7 Guns are sometimes used: I used a starting pistol in the Inchindown Oil Tank when measuring the world record reverberation. Cox, T., 2014. *Sonic Wonderland: A Scientific Odyssey of Sound*. Random House.

8 She also placed the sounds in slightly different positions.

9 An excellent discussion of vocal staging is Lacasse, S., 2000. 'Listen to my voice. the evocative power of voice in recorded rock music and other forms of vocal expression'. PhD thesis, University of Liverpool, where other examples of flat reproduction can be found.

10 A more complete scientific study involving brain scanning and perceptual measurements is Kumar, S., von Kriegstein, K., Friston, K. and Griffiths, T. D., 2012. 'Features versus feelings: dissociable representations of the acoustic features and valence of aversive sounds'. *Journal of Neuroscience*, 32(41), pp. 14184–92.

11 Arnal, L. H., Flinker, A., Kleinschmidt, A., Giraud, A. L. and Poeppel, D., 2015. 'Human screams occupy a privileged niche in the communication soundscape'. *Current Biology*, 25(15), pp. 2051–6.

12 LeDoux, J. E., 2015. 'The Amygdala Is NOT the Brain's Fear Center'. www.psychologytoday.com/blog/i-got-mind-tell-you/201508/the-amygdala-is-not-the-brains-fear-center, accessed 18 March 2017.

13 The S3A project involves Surrey, Salford and Southampton universities, as well as BBC R&D.

14 Weaver, M., 2017. '"I will mumble this only once": BBC's Nazi drama *SS-GB* hit by dialogue complaints'. *Guardian*.

15 Gentleman, A. and Gibbons, F., 1999. 'Outcry at Nunn's use of mikes in theatre'. *Guardian*.

16 Billington, M., 1999. 'Review: *Troilus and Cressida*'. Guardian.

17 He has also twice won the Laurence Olivier Award for Best Sound Design.

18 Adding reverberation can also be used to signal more than just room size. When I talked to Eloise Whitmore about radio drama, she explained how reverberation is often a signal of a flashback or something not quite real.

19 Cox, T., 2014. 'Reverb: Why we dig messy sound'. *New Scientist*, 3000.

20 Another example of music producers playing on expectation of listeners is Daniel Lanois' *Death of a Train* from 1993. The slap-back echo on this track harks back to the type of processing pioneered by Sun Records back in the 1950s. This then adds an air of nostalgia to Lanois' song.

21 Fry G., 2015. 'Capturing sound for Complicité's *The Encounter*'. *Stage*. www.thestage.co.uk/features/2015/backstage-a-trip-to-the-amazon-rainforest-to-capture-the-perfect-sound/, accessed 18 March 2017.

22 Chapter 1 described these cues.

23 While the need for projection has hugely shaped the operatic voice, it is important not to overlook other cultural factors, see Potter, op. cit.

24 In opera, the audience often cannot understand the language being sung anyway. Freddie Mercury and Montserrat Caballé. 'Barcelona'. https://youtu.be/hkskujG0Uyc, accessed 18 March 2017.

25 According to John Potter the current operatic voice is probably different from that in the nineteenth century. Contemporaneous reports indicate that a wider range of singing styles was probably used.

26 Long, C., 2016. 'Bob Dylan and the Manchester Free Trade Hall "Judas" show'. BBC. www.bbc.co.uk/news/entertainment-arts-36211789, accessed 18 March 2017.

27 They also change the vocal fold vibration to amplify the harmonics in the ear's most sensitive bandwidth. There is still much debate about whether actresses use formant tuning. Latest evidence suggests they probably do not and just rely on changing the vocal fold vibration. Master, S., De Biase, N. G. and Madureira, S., 2012. 'What About the "Actor's Formant" in Actresses' Voices?' *Journal of Voice*, 26(3), pp. e117–e122.

28 Smith, J. and Wolfe, J., 2009. 'Vowel-pitch matching in Wagner's operas: implications for intelligibility and ease of singing'. *JASA Express Letters, Journal of the Acoustical Society of America*, 125, EL196.

29 Al Bowlly. 'Melancholy Baby'. https://youtu.be/rRF5D68e44Q, accessed 18 March 2017. This example came from Potter, J. and Sorrell, N., 2012. *A History of Singing*. Cambridge University Press.

30 Frith, S., 1986. 'Art versus technology: The strange case of popular music'. *Media, Culture & Society*, 8(3), pp. 263–79.

31 Leslie, J. and Snyder, R., 2010. *History of The Early Days of Ampex Corporation*. AES Historical Committee.

32 Hammer, P., 1994. 'In Memoriam John T. (Jack) Mullin'. *Journal of the Audio Engineering Society*, 42(6).

33 'Billie Holiday Dies Here at 44; Jazz Singer Had Wide Influence'. *New York Times*, 1959. www.nytimes.com/learning/general/onthisday/bday/0407.html, accessed 18 March 2017.

34 Empire, K. 2006. 'Let's judge women on their talent, not their pain'. *Guardian*.

35 Adele. 'Someone Like You'. https://youtu.be/hLQl3WQQoQ0, accessed 18 March 2017.

36 Compression actually reduces the level of loud parts of the music but is done in tandem with turning up the volume to end up with boosting the quiet parts.
37 Fashions may be changing, however, with breathy voices becoming more common. Robinson, P., 2017. '"Whisperpop": why stars are choosing breathy intensity over vocal paint-stripping'. *Guardian.*
38 McCormick, N., 2005. 'Take that, says Robbie as he faces his critics'. *Telegraph.*
39 Quotes from BBC Radio 4, 2014. 'Creating Pitch-Perfect'. http://www.bbc.co.uk/programmes/b01shwkq, accessed 4 March 2014.
40 Dan also said how you have to balance the rhythm so you do not end up over-inflating or deflating.
41 Rahzel. 'If Your Mother Only Knew'. https://youtu.be/ifCwPidxsqA, accessed 18 March 2017.
42 Unless they just want to perform a cappella, beatboxers use a microphone to get more bass from their voice. With most vocal microphones, as you get it closer to your mouth you get more bass due to *the proximity effect.*
43 If you listen to Rahzel in detail, you'll notice the 'l' in 'only' is often replaced by a compound sound.
44 *Sparky and the Talking Train.* https://www.youtube.com/watch?v=3O3IzIzoV-vk&feature=youtu.be, accessed 18 March 2017.
45 The actor needs to close the glottis, otherwise a lot of the sound disappears into the lungs.
46 Make sure you listen to the album version. https://youtu.be/x-G28iyPtz0, accessed 18 March 2017.
47 Hildebrand, H. A., Auburn Audio Technologies Inc., 1999. 'Pitch detection and intonation correction apparatus and method'. US Patent 5,973,252.
48 These is a great auditory illusion called 'Speech to Song' by Diana Deutsch that exploits this. deutsch.ucsd.edu/psychology/pages.php?i=212, accessed 18 March 2017.
49 Another example comes from the shifting voices on 'A Day in the Life' by the Beatles, with the effect being most obvious on headphones. When Lennon starts singing 'I read the news today' his voice is all the way to the right. By the time the first verse is ending with 'I'd love to turn you on' he has drifted all the way to the left. This has echoes of the age-old tradition of antiphonal singing, where two separated choirs sing in call and response to spatially distribute the music.

Chapter 6

1 'An Evening With Edison'. *New York Times*, 6 April 1878.
2 Preece, W. H., 1878. 'The Phonograph'. *Journal of the Society of Arts*, 26, p. 537.
3 Thompson, E., 1995. 'Machines, Music, and the Quest for Fidelity: Marketing the Edison Phonograph in America, 1877–1925'. *Musical Quarterly*, 79(1), pp. 131–71.
4 http://www.bbc.co.uk/mediacentre/latestnews/2016/bbc-russian-virtual-voice-over, accessed 8 January 2017.
5 Dudley, H. and Tarnoczy, T. H., 1950. 'The speaking machine of Wolfgang von Kempelen'. *Journal of the Acoustical Society of America*, 22(2), pp. 151–66.
6 Davis, A., 2016. 'Mechanical chess player baffled crowds for nearly a century'. IEEE, accessed 12 November 2016.
7 Hear a demo: https://youtu.be/k_YUB_S6Gpo. The machine is different from Howard's, but the principles are the same. Brackhane, F. and Trouvain, J., 2008. 'What makes "Mama" and "Papa" acceptable? – Experiments with a replica of von Kempelen's speaking machine'. *Proceedings of the 8th International Speech Production Seminar*, pp. 333–6.
8 Translation from Trouvain, J. and Brackhane, F., 2011. 'Wolfgang von Kempelen's "speaking machine" as an instrument for demonstration and research' in W.-S. Lee and E. Zee (eds.), *Proceedings of the 17th International Congress of Phonetic Sciences*, pp. 164–7.
9 'The Speaking Machine'. Punch, 11 (1846), p. 83.
10 Altick, R. D., 1978. *The Shows of London*. Harvard University Press, p. 355.
11 'The New York Fair'. *Bell Telephone Quarterly*, January 1940, p. 63.
12 Schroeder, M. R., 1981. 'Dudley, Homer W.: A Tribute'. *Journal of the Acoustical Society of America*, 69(4), p. 1222.
13 Dlugan, A., 2012. 'What is the Average Speaking Rate?'. http://sixminutes.dlugan.com/speaking-rate/, accessed 18 April 2017.
14 Drawing based on Dudley, H., Riesz, R. R and Watkins, S. S. A., 1939. 'A synthetic speaker'. *Journal of the Franklin Institute*, 227, pp. 739–64.
15 Historical recordings can be found online, e.g. https://youtu.be/5hyI_dM5cGo. The later quote is from this video.
16 Fagen, M. D., Millman, S., Joel, A. E. and Schindler, G. E., 1975. *A History of Engineering and Science in the Bell System: Communications sciences (1925–1980), Vol. 5*. Bell Telephone Laboratories Inc., pp. 101ff.

17 http://www.ti.com/corp/docs/company/history/timeline/eps/1970/docs/78-speak-spell_introduced.htm, accessed 18 April 2017.

18 'Who is Hatsune Miku?'. http://www.crypton.co.jp/miku_eng, accessed 18 April 2017.

19 Kenmochi, H. and Ohshita, H., 2007. 'VOCALOID-commercial singing synthesizer based on sample concatenation'. *Interspeech*, pp. 4009–10.

20 https://youtu.be/UQw03TXZsHA, accessed 19 April 2017. Unfortunately the sound quality on the video is poor.

21 Boone, J. V. and Peterson, R. R., 2016. 'Sigsaly – The Start of the Digital Revolution'. https://www.nsa.gov/about/cryptologic-heritage/historical-figures-publications/publications/wwii/sigsaly-start-digital.shtml, accessed 18 April 2017.

22 Kahn, D., 2014. *How I Discovered World War II's Greatest Spy and Other Stories of Intelligence and Code*. CRC Press.

23 99% Invisible Podcast, 2016. 'Vox Ex Machina'. http://99percentinvisible.org/episode/vox-ex-machina/, accessed 18 April 2017.

24 http://www.acapela-group. com/voices/demo/, accessed 18 April 2017.

25 Victoria T., 2016. 'How we fell in love with our voice-activated home assistants'. *New Scientist*, 3104.

26 Parke, P, 2015. 'Is it cruel to kick a robot dog?'. CNN, http://edition.cnn.com/2015/02/13/tech/spot-robot-dog-google/index.html, accessed 27 November 2017.

27 See demos at https://www.biomotionlab.ca/. Also Waytz, A., Epley, N. and Cacioppo, J. T., 2010. 'Social cognition unbound: Insights into anthropomorphism and dehumanization'. *Current Directions in Psychological Science*, 19(1), pp. 58–62.

28 Newman, J., 2014. 'To Siri with Love: How One Boy With Autism Became BFF With Apple's Siri'. *New York Times*.

29 Ramaswamy, C., 2017. '"Alexa, sort your life out": when Amazon Echo goes rogue'. *Guardian*. In 2015, Samsung got unwanted headlines when it was revealed that recordings being picked up by its smart TV remote was being sent to a third-party company that was carrying out the speech analysis.

30 Hern, A., 2017. 'Murder defendant volunteers Echo recordings Amazon fought to protect'. *Guardian*.

31 Eng, J., 2016. 'NYC to Parents: Make Sure Your Baby Monitors Don't Get Hacked'. NBC News.

32 Walker, T, 2017. 'How local accents have replaced Stephen Hawking-style voice boxes'. *Guardian.*

33 They tend to average across donor voices because it gives a better end result.

34 Play with demos at: http://www.nutbot.net/talking_head/, accessed 10 September 2017.

35 The researchers also got people talking in regional and foreign accents. McGettigan, C., Eisner, F., Agnew, Z. K., Manly, T., Wisbey, D. and Scott, S. K., 2013. 'T'ain't what you say, it's the way that you say it – left insula and inferior frontal cortex work in interaction with superior temporal regions to control the performance of vocal impersonations'. *Journal of Cognitive Neuroscience*, 25(11), pp. 1875–86.

36 The Naked Scientists, 2016. 'The neuroscience of a good impression'. www.thenakedscientists.com/articles/interviews/neuroscience-good-impression, accessed 27 November 2017.

37 Logan, T., 2007. 'Nice talking to you, machine'. *New Scientist*, 2590.

38 Spinney, L., 2017. 'Exploring the uncanny valley: Why almost-human is creepy'. *New Scientist*, 3097.

39 Mäkäräinen, M., Kätsyri, J., Förger, K. and Takala, T., 2015. 'The funcanny valley: A study of positive emotional reactions to strangeness'. *Proceedings of the 19th International Academic Mindtrek Conference*, pp. 175–81.

40 Kätsyri, J., Förger, K., Mäkäräinen, M. and Takala, T., 2015. 'A review of empirical evidence on different uncanny valley hypotheses: Support for perceptual mismatch as one road to the valley of eeriness'. *Frontiers in Psychology*, 6, p. 390.

41 Mitchell, W. J., Szerszen Sr, K. A., Lu, A. S., Schermerhorn, P. W., Scheutz, M. and MacDorman, K. F., 2011. 'A mismatch in the human realism of face and voice produces an uncanny valley'. *i-Perception, 2*(1), pp. 10–12. Tinwell, A., Grimshaw, M. and Nabi, D. A., 2015. 'The effect of onset asynchrony in audio-visual speech and the Uncanny Valley in virtual characters'. *International Journal of Mechanisms and Robotic Systems*, 2(2), pp. 97–110.

42 https://www.lifenaut.com/bina48/, accessed 18 April 2017.

43 Bina48 also lacks what linguists call backchannels, those little responses we all make to show we are paying attention, like *Fawlty Towers*' Sybil's 'Oh I knowww.'

44 Another hypothesis is that it arises from the disgust reaction. A humanoid that does not look quite right somehow appears wrong to a human, for example seeming diseased, and therefore we have a revulsion reaction to keep us away.

45 This sentiment was also echoed by Judy Norman who acted alongside a robot in the play *Spillikin: A Love Story*. She commented in a Q&A to the audience that the robot was 'not that much different from an actor to tell the truth, I thought it was going to be radically different, but actually it's not really.' She also commented that the 'most difficult thing is the robot can't help you get out of any mistake you make'.

46 Using a human to speak the lines of a computer character is normal: in the film *2001: A Space Odyssey*, Hal is voiced by Canadian actor Douglas Rain.

47 It would also have been possible to use some automatic features, for example the robot could try to make eye contact.

Chapter 7

1 More details of the case can be found at http://news.bbc.co.uk/1/hi/world/americas/3243015.stm, accessed 22 April 2015, and in Levi-Minzi, M. and Shields, M., 2007. 'Serial sexual murderers and prostitutes as their victims: Difficulty profiling perpetrators and victim vulnerability as illustrated by the Green River case'. *Brief Treatment and Crisis Intervention*, 7(1), p. 77.

2 Party, B. W., 2004. 'A review of the current scientific status and fields of application of polygraphic deception detection'. *British Psychological Society.*

3 'Innocent Until Proved Guilty?' ABC News, 2006. http://abcnews.go.com/Primetime/story?id=1786421, accessed 22 April 2015.

4 Juslin, P. N. and Laukka, P., 2003. 'Communication of emotions in vocal expression and music performance: Different channels, same code?' *Psychological Bulletin*, 129(5), p. 770.

5 The published work on theme tunes used support vector machines rather than the artificial neural networks because we were curious to see how well they would do. Mann, M., Cox, T. J. and Li, F. F., 2011. 'Music Mood Classification of Television Theme Tunes'. *International Society of Music Information Retrieval*, pp. 735–40.

6 If you have large amounts of data and sufficient computing grunt, then you can feed the raw audio into a Deep Neural Network and skip the feature extraction part. But we did not have large amounts of data.

7 You can hear the tune at: http://www.televisiontunes.com/Noggin_the_Nog.html, accessed 24 April 2017.

8 Kreiman and Sidtis, op. cit.

9 Elaad, E., 2003. 'Effects of feedback on the overestimated capacity to detect lies and the underestimated ability to tell lies'. *Applied Cognitive Psychology*, 17(3), pp. 349–63.

10 Another tactic is to appeal to the teenager's competitive instincts by telling them that they are trying to deceive highly gifted lie catchers. Actors are best avoided as they are likely to present exaggerated, stereotypical cues.

11 The average accuracy across studies is 54 per cent; Bond, C. F. and DePaulo, B. M., 2006. 'Accuracy of deception judgments'. *Personality and social psychology Review*, 10(3), pp. 214–34. There is disagreement between deception researchers about whether there are *Wizards* who are particularly good at spotting lies through micro-expressions. This is the premise used in the TV series *Lie to Me*.

12 Wiseman, R., 1995. 'The megalab truth test'. *Nature*, 373(6513), p. 391.

13 This was true in 90 per cent of countries. Vrij, A., Granhag, P. A. and Porter, S., 2010. 'Pitfalls and opportunities in nonverbal and verbal lie detection'. *Psychological Science in the Public Interest*, 11(3), pp. 89–121.

14 Shermer, M., 2011. *The Believing Brain*. Macmillan.

15 Unfortunately, one group that gets immediate feedback on their lying behaviour is career criminals, allowing them to learn countermeasures to fool their interrogators.

16 Liars are also concentrating on presenting a credible story and there is an additional cognitive load if you are constantly checking what impression others might be taking from a statement. This could also be true of nervous truth-tellers, however.

17 Vrij, A., Edward, K. and Bull, R., 2001. 'People's insight into their own behaviour and speech content while lying'. *British Journal of Psychology*, 92(2), pp. 373–89.

18 Serota, K. B., Levine, T. R. and Boster, F. J., 2010. 'The Prevalence of Lying in America: Three Studies of Self-Reported Lies'. *Human Communication Research*, 36(1), pp. 2–25.

19 Ten Brinke, L., Stimson, D. and Carney, D. R., 2014. 'Some evidence for unconscious lie detection'. *Psychological Science*, p. 0956797614524421.

20 We met this test method previously when looking at vocal prejudice.

21 BBC News, 2003. http://news.bbc.co.uk/1/hi/uk/3227849.stm, accessed 22 April 2015.

22 Heingartner, D., 2004. 'It's the Way You Say It, Truth Be Told'. *New York Times*. http://www.nytimes.com/2004/07/01/technology/it-s-the-way-you-say-it-truth-be-told.html, accessed 29 April 2017.

23 Lacerda, F., June 2009. 'LVA technology: The illusion of lie detection'. *FONETIK.*

24 Kreiman and Sidtis, op. cit., p. 369.

25 Eriksson, A. and Lacerda, F., 2007. 'Charlatanry in forensic speech science: A problem to be taken seriously'. *International Journal of Speech, Language and the Law,* 14(2), pp. 169–93.

26 BBC News, 2013. 'Defamation Act 2013 aims to improve libel laws'. http://www.bbc.co.uk/news/uk-25551640, accessed 22 April 2015.

27 Damphousse, K. R., Pointon, L., Upchurch, D. and Moore, R. K., 2007. 'Assessing the validity of voice stress analysis tools in a jail setting'. Report submitted to the US Department of Justice.

28 When the arrestees thought their speech was being analysed, 14 per cent lied about recent drug use, whereas 40 per cent lied when they were unaware of the voice stress analysis. Damphousse et al., op. cit.

29 Jones, E. E. and Sigall, H., 1971. 'The bogus pipeline: a new paradigm for measuring affect and attitude'. *Psychological Bulletin,* 76(5), p. 349.

30 Arthur, C., 2009. 'Government data shows £2.4m "lie detection" didn't work in 4 of 7 trials' and http://www.ministryoftruth.me.uk/2012/02/08/nemesyscos-lva-technology-ghosts-in-the-noise/#disqus_thread, accessed 22 April 2015.

31 Ekman, P., 2009. *Telling Lies: Clues to Deceit in the Marketplace, Politics, and Marriage* (revised edition). WW Norton & Company.

32 BBC News, 1989. 'Exxon Valdez creates oil slick disaster'. http://news.bbc.co.uk/onthisday/hi/dates/stories/march/24/newsid_4231000/4231971.stm, accessed 22 April 2015.

33 Schuller, B., Batliner, A., Steidl, S., Schiel, F. and Krajewski, J., January 2011. 'The Interspeech 2011 speaker state challenge'. *Interspeech,* pp. 3201–4.

34 Bone, D., Black, M., Li, M., Metallinou, A., Lee, S. and Narayanan, S. S., August 2011. 'Intoxicated Speech Detection by Fusion of Speaker Normalized Hierarchical Features and GMM Supervectors'. *Interspeech,* pp. 3217–20.

35 Success rate is 74 per cent when two samples from the same speaker, one recorded while the person is intoxicated and one while sober, are compared. If just presented with one sound sample, so no comparison is possible, then success rates drop to 65 per cent. Pisoni, D. B. and Martin, C. S., 1989. 'Effects of Alcohol on the Acoustic-Phonetic Properties of Speech: Perceptual and Acoustic Analyses'. *Alcoholism: Clinical and Experimental Research,* 13(4), pp. 577–87.

36 Emotion, fatigue or stress could also have had an effect. Kreiman and Sidtis, op. cit., p. 360.

37 Oberlader, V. A., Naefgen, C., Koppehele-Gossel, J., Quinten, L., Banse, R. and Schmidt, A. F., 2016. 'Validity of content-based techniques to distinguish true and fabricated statements: A meta-analysis'. *Law and Human Behavior*, 40(4), p. 440.

38 Jim Flanagan et al., 1980. 'Techniques for expanding the capabilities of practical speech recognizers'. *Trends in Speech Recognition*. For more on Audrey: Davis, K. H., Biddulph, R. and Balashek, S., 1952. 'Automatic recognition of spoken digits'. *Journal of the Acoustical Society of America*, 24(6), pp. 637–42.

39 BBC, 2011. 'Say what? iPhone has problems with Scots accents'. http://www.bbc.co.uk/news/uk-scotland-15475989, accessed 22 April 2015.

40 Caliskan, A., Bryson, J. J. and Narayanan, A., 2017. 'Semantics derived automatically from language corpora contain human-like biases'. *Science*, 356(6334), pp. 183–6.

41 This example came from *Science* News, 2017. 'Biased bots: Human prejudices sneak into artificial intelligence systems'. https://www.sciencedaily.com/releases/2017/04/170413141055.htm, accessed 25 April 2017.

42 Dahl, G. E., Yu, D., Deng, L. and Acero, A., 2012. 'Context-dependent pre-trained deep neural networks for large-vocabulary speech recognition'. *Audio, Speech, and Language Processing, IEEE Transactions*, 20(1), pp. 30–42.

43 Rayner, K., White, S. J., Johnson, R. L. and Liversedge, S. P., 2006. 'Raeding Wrods With Jubmled Lettres There Is a Cost'. *Psychological science*, 17(3), pp. 192–3.

44 In 2011, Google Search by Voice was trained on 240 billion words from millions of users. 'Speech Recognition Lightning Talk – Google and AAAI 2011'. https://www.youtube.com/watch?v=g6iAOdRsDOM, accessed 22 April 2015.

45 Dong, X. L., Gabrilovich, E., Murphy, K., Dang, V., Horn, W., Lugaresi, C., Sun, S. and Zhang, W., 2016. 'Knowledge-Based Trust: Estimating the Trustworthiness of Web Sources'. *IEEE Data Eng. Bulletin*, 39(2), pp. 106–17.

46 Chilton, M., 2015. 'The best spoonerisms'. *Telegraph.*

47 Kiddon, C. and Brun, Y., 2011. 'That's what she said: double entendre identification'. *Proceedings of the 49th Annual Meeting of the Association for Computational Linguistics: Human Language Technologies*, 2. pp. 89–94.

48 Scott, S. K., Lavan, N., Chen, S. and McGettigan, C., 2014. 'The social life of laughter'. *Trends in Cognitive Sciences*, 18(12), pp. 618–20.

49 Involuntary laughter has longer total duration, shorter bursts, higher pitch, more unvoiced segments and lower mean intensity than posed laughter. Posed laughter has more nasality. Lavan, N., Scott, S. K. and McGettigan, C., 2016. 'Laugh like you mean it: Authenticity modulates acoustic, physiological and perceptual properties of laughter'. *Journal of Nonverbal Behavior*, 40(2), pp. 133–49.

50 Carr tells a great anecdote about his time watching fellow comedian Nick Helm perform. As Carr explained, 'He did a routine that I found hilarious and I had a proper full-on laughing fit. Nick just stopped the show and went, "All right Jimmy, I don't laugh when I come to see one of your shows."' http://www.digitalspy.com/tv/news/a788818/jimmy-carr-compares-laugh-weird-honking-goose-while-talking-new-netflix-show/, accessed 27 April 2017.

51 Crying is similarly distinctive and also increases activity in the auditory cortex; see Arnal, L. H., Flinker, A., Kleinschmidt, A., Giraud, A. L. and Poeppel, D., 2015. 'Human screams occupy a privileged niche in the communication soundscape'. *Current Biology*, 25(15), pp. 2051–6.

52 Truong, K. P. and Van Leeuwen, D. A., 2007. 'Automatic discrimination between laughter and speech'. *Speech Communication*, 49(2), pp. 144–58.

53 http://www.imdb.com/character/ch0030014/quotes, accessed 22 April 2015. Quote from Lynn, J. and Jay, A., 1984. *The Complete Yes Minister: The Diaries of a Cabinet Minister, by the Right Hon. James Hacker MP*. BBC Books.

54 BBC World Service. *The Why Factor*, 'The Lie'. http://www.bbc.co.uk/programmes/p0188dbq, broadcast 20 May 2013.

55 McNally, L. and Jackson, A. L., July 2013. 'Cooperation creates selection for tactical deception'. *Proceedings of the Royal Society B*, 280(1762), p. 20130699. The Royal Society.

Chapter 8

1 According to Barak Turovsky, head of product management and user experience at Google Translate. https://www.theguardian.com/technology/2016/jun/04/man-v-machine-robots-artificial-intelligence-cook-write, accessed 25 April 2017.

2 Campbell-Kelly, M., 1980. 'Programming the Mark I: Early programming activity at the University of Manchester'. *Annals of the History of Computing*, 2(2), pp. 130–68.

3 https://transmediale.de/content/there-must-be-an-angel-on-the-beginnings-of-arithmetics-of-rays, accessed 26 May 2017.

4 http://jamesrobertlloyd.com/blog-2016–04–18–poetry-net, accessed 30 April 2016.

5 This idea was a thought experiment by Turing, rather than a definite proposal for a test. Turing, A. M., 1950. 'Computing machinery and intelligence'. *Mind*, 59(236), pp. 433–60.

6 Jefferson, G., 1949. 'The Mind of Mechanical Man'. *British Medical Journal*, 1(4616), pp. 1105–10.

7 Misztal-Radecka, J. and Indurkhya, B., 2016. 'A blackboard system for generating poetry'. *Computer Science*, 17(2), p. 265.

8 http://hyperboleandahalf.blogspot.co.uk/2013/05/depression-part-two.html, accessed 28 April 2017.

9 Steinbeis, N. and Koelsch, S., 2009. 'Understanding the intentions behind man-made products elicits neural activity in areas dedicated to mental state attribution'. *Cerebral Cortex*, 19(3), pp. 619–23. Another study in this area found ratings of music decreased when people were told they were written by computer. Unsurprisingly, the prejudice against computer compositions was most marked among musicians. Moffat, D. C. and Kelly, M., 2006. 'An investigation into people's bias against computational creativity in music composition'. *Proceedings of the 3rd International Joint Workshop on Computational Creativity*. ECAI06 Workshop, Riva del Garda, Italy.

10 https://github.com/thricedotted/theseeker/blob/master/the_seeker.pdf, accessed 25 April 2017.

11 White, E. M., 2015. 'Automated earnings stories multiply'. AP. https://blog.ap.org/announcements/automated-earnings-stories-multiply. See also Benedictus, L., 2016. 'Man v machine: can computers cook, write and paint better than us?' *Guardian*; and AP, 'AP expands Minor League Baseball coverage', https://www.ap. org/press-releases/2016/ap-expands-minor-league-baseball-coverage, accessed 8 October 2017.

12 Riedl, M. O., 2016. 'Computational Narrative Intelligence: A Human-Centered Goal for Artificial Intelligence'. arXiv preprint arXiv:1602.06484.

13 You can hear GenJam play at https://www.youtube.com/watch?v=rF-BhwQUZGxg, accessed 2 May 2017.

14 Boden, M. A., 2004. *The Creative Mind: Myths and Mechanisms*. Psychology Press, pp. 1–10.

15 In 2003 Paignton Zoo tried this with six Sulawesi crested macaque monkeys. The five pages of text they got were mostly just filled with the letter 's'. Adam,

D., 2003. 'Give six monkeys a computer, and what do you get? Certainly not the Bard'. *Guardian*.

16 Computers that create puns and other wordplay jokes are using the same sort of creativity. Ritchie, G., 2009. 'Can Computers Create Humor?' *AI Magazine*, 30(3), pp. 71–81.

17 A short sci-fi movie has also been made using artificial intelligence. Newitz, A., 2016. 'Movie written by AI algorithm turns out to be hilarious and intense'. https://arstechnica.co.uk/the-multiverse/2016/06/sunspring-movie-watch-written-by-ai-details-interview/, accessed 28 April 2017.

18 Llano, M. T., Colton, S., Hepworth, R. and Gow, J., 2016. 'Automated fictional ideation via knowledge base manipulation'. *Cognitive Computation*, 8(2), pp. 153–74.

19 Parkin, S., 2014. 'Automatic authors: Making machines that tell tales'. *New Scientist*, 2990.

20 Schwarm, B., 2013. 'The Hebrides, Op. 26'. http://www.britannica.com/topic/The-Hebrides-Op-26, accessed 25 April 2017.

21 The fuller narrative arc was constructed using PropperWryter using research led by Dr Pablo Gervás, University of Madrid. Colton, S., Llano, M. T., Hepworth, R., Charnley, J., Gale, C. V., Baron, A., Pachet, F., Roy, P., Gervás, P., Collins, N. and Sturm, B., June 2016. 'The beyond the fence musical and computer says show documentary'. *Proceedings of the International Conference on Computational Creativity.*

22 Accidentally, some show reviews were included in the data, so while it mostly produced a stream of poetry, every so often the neural network would suddenly switch to impersonating an illiterate critic. One section of output was, 'The cast melody and the dance musical show can seem to fear the thrill of the popular and analyzing favor of a hit which the production is the 2004 version of The Boat of the Party and The Indian show.'

23 People are attempting lyric generation, e.g. Gonçalo Oliveira, H., 2015. 'Tra-la-lyrics 2.0: Automatic generation of song lyrics on a semantic domain'. *Journal of Artificial General Intelligence*, 6(1), pp. 87–110.

24 Ghedini, F., Pachet, F. and Roy, P., 2016. 'Creating music and texts with flow machines'. *Multidisciplinary Contributions to the Science of Creative Thinking.* Springer Singapore, pp. 325–43.

25 More recently Microsoft and the University of Cambridge have produced a system to write very short programs by harvesting online code. The idea is to allow people to describe an idea for a program and then let the system

write the code. Balog, M., Gaunt, A. L., Brockschmidt, M., Nowozin, S. and Tarlow, D., 2016. 'DeepCoder: Learning to Write Programs'. arXiv preprint arXiv:1611.01989.

26 Metz, C., 2016. 'AI is transforming Google search. The rest of the web is next'. *Wired.* http://www.wired.com/2016/02/ai-is-changing-the-technology-behind-google-searches/, accessed 25 April 2017.

27 Cox, T. J. and D'Antonio, P., 2016. *Acoustic Absorbers and Diffusers: Theory, Design and Application.* CRC Press.

28 Riedl, M. O., 2014. 'The Lovelace 2.0 Test of Artificial Creativity and Intelligence'. arXiv preprint arXiv:1410.6142, accessed 8 December 2017.

29 Steadman, I., 2013. 'IBM's Watson is better at diagnosing cancer than human doctors'. *Wired.* http://www.wired.co.uk/news/archive/2013–02/11/ibm-watson-medical-doctor. It is claimed that this has reached the knowledge of the average second-year medical student. IBM, 2014. 'IBM Watson Ushers in a New Era of Data-Driven Discoveries'. https://www-03.ibm.com/press/us/en/pressrelease/44697.wss, accessed 25 April 2017.

30 It can also remove experimenter bias, which is one of the reasons computers are starting to impact on neuroscience. Lorenz, R., Monti, R. P., Violante, I. R., Anagnostopoulos, C., Faisal, A. A., Montana, G. and Leech, R., 2016. 'The automatic neuroscientist: a framework for optimizing experimental design with closed-loop real-time fMRI'. *NeuroImage*, 129, pp. 320–34.

31 'Pullman, Philip'. *Oxford Encyclopedia of Children's Literature*, 2006.

32 Dijksterhuis, A., Bos, M. W., Nordgren, L. F. and Van Baaren, R. B., 2006. 'On making the right choice: the deliberation-without-attention effect'. *Science*, 311, pp. 1005–7.

33 And up to much higher levels of abstraction such as anticipating how others might respond to the poem and guessing who might be next to try the machine.

34 It is not just the preserve of the right hemisphere as the popular myth would suggest.

Index

Author photograph courtesy of the University of Salford

TREVOR COX is professor of acoustic engineering at the University of Salford, U.K., and a former president of the Institute of Acoustics. He has presented twenty-four science documentaries on BBC Radio and written feature articles for *New Scientist*, *Sound on Sound*, and *The Guardian*. He is the author of *The Sound Book*, winner of an Acoustical Society of America writing award. He was awarded the Institute of Acoustics' Tyndall Medal and the Institute of Acoustics' Award for Promoting Acoustics to the Public, and currently holds the Guinness World Record for producing the longest echo in one of the Inchindown Oil Tanks.